Energy Systems in Electrical Engineering

Series Editor

Muhammad H. Rashid, Florida Polytechnic University, Lakeland, USA

More information about this series at http://www.springer.com/series/13509

Qing-Hua Wu · Jiehui Zheng ·
Zhaoxia Jing · Xiaoxin Zhou

Large-Scale Integrated Energy Systems

Planning and Operation

Springer

Qing-Hua Wu
School of Electric Power Engineering
South China University of Technology
Guangzhou, China

Jiehui Zheng
School of Electric Power Engineering
South China University of Technology
Guangzhou, China

Zhaoxia Jing
School of Electric Power Engineering
South China University of Technology
Guangzhou, China

Xiaoxin Zhou
China Electric Power Research Institute
Beijing, China

ISSN 2199-8582 ISSN 2199-8590 (electronic)
Energy Systems in Electrical Engineering
ISBN 978-981-13-6942-1 ISBN 978-981-13-6943-8 (eBook)
https://doi.org/10.1007/978-981-13-6943-8

This Springer imprint is published by the registered company Springer Nature Singapore Pte Ltd.
The registered company address is: 152 Beach Road, #21-01/04 Gateway East, Singapore 189721, Singapore

To our families

Preface

Integrated energy systems are introduced to address efficient utilization of various distributed energy resources in order to satisfy diverse types of energy demands, such as electrical loads, heating loads, and cooling loads. The integrated energy system is typically a complex system with the characteristics of heterogeneity, hierarchical structure, complex network topology, multi-modal, hybrid variables, and complex dynamics.

In recent years, there has been numerous research projects on the design, control, and operation of the integrated energy systems, such as the CERT microgrid in USA, the European MICROGRID project, and the new energy integration test project in Japan. Besides, many researches have focused on the uncertainties of renewable resources penetrated into power grids and the optimal operation of microgrids. Those approaches usually focus on part of the integrated energy system without considering the synergies between different operating interests. However, as the types of renewable energy resources and energy demands increase, it is of great importance to integrate the distributed energy systems and the power grid in the integrated energy system in order to satisfy not only the electricity loads but also the heating loads and cooling loads.

This book is dedicated to research on the core fundamental technologies of integrated energy system planning and operation, which consist of three parts: (1) Modeling of integrated energy systems, including modeling of the dynamic behaviors of various energy devices as well as their connections to form an integrated energy system model; (2) development of high-dimensional multi-objective stochastic optimization algorithms, and (3) development of decision-making systems for determination of the final optimal solution of planning and operation of integrated energy systems, selected from the Pareto sets of the multi-objective optimization computation. The developed algorithms and methods are applied to construct a power network based integrated energy system and investigate the economy and reliability of integrated energy systems which could be achieved using distributed CHP and CCHP, heat storages, and cool storages, and investigate the smooth peaks and valleys of power generation and loads, respectively.

We wrote this book in the belief that planning and operation problems of integrated energy systems are critical to the integrated utilization of various energies such as coal, wind power, solar energy, nuclear power, and hydropower for thermal, cool and electric loads. This book is self-contained with adequate background introductions underlying analytical solutions of the planning and operation problems of integrated energy systems and links to the publicly available toolboxes for the implementation of the modeling, optimization, and decision-making methodologies. This book presents many examples of using benchmark systems and realistic operating scenarios of integrated energy systems, which fully illustrate the core fundamental technologies of modeling, optimization, and decision-making to deal with the planning and operation problems of integrated energy systems. We hope that this book will be useful for those postgraduates, academic researchers, and engineers working in the area of integrated energy system planning and operation.

We would like to thank Muhammad H. Rashid, Umamagesh A. P. and Megana Dinesh, the editors of Springer, for their professional and efficient editorial work on this book. Our thanks are also extended to all colleagues in the Smart Grid and Its Automation Group, South China University of Technology, for all assistance provided, and who have not been specially mentioned above.

This book is based on the work partly funded by the State Key Program of National Natural Science Foundation of China (Grant No. 51437006).

Guangzhou, China
March 2019

Prof. Qing-Hua Wu
Dr. Jiehui Zheng

Contents

Acronyms

ABC	Artificial bee colony
AC	Absorption chiller
ACO	Ant colony optimization
ACS	Artificial cooperative search
AHU	Air handling unit
AP	Affinity propagation
BB	Branch and bound
BBO	Biogeography-based optimization
BSA	Backtracking search optimization algorithm
CCHP	Combined cooling, heating and power
CDCENs	Coupling demand constraints of electricity and natural gas
CDE	Carbon dioxide emissions
CEED	Combined economic emission dispatch
CK	Cuckoo-search algorithm
CLPSO	Comprehensive learning particle swarm optimizer
CLTD	Cooling load temperature difference
CMA-ES	Covariance matrix adaptation evolution strategy
COP	Coefficient of performance
CS	Cooling system
CSS	Coordinated scheduling strategy
CT	Computation time
DEN	District energy network
DESs	Distributed energy stations
DH	District heating system
DHCs	District heating and cooling units
DSA	Differential search algorithm
E&C	Electricity and cooling
E&G	Electricity and natural gas
EAs	Evolutionary algorithms
EB	Electric water boiler

EC	Electric chiller
ED	Economic dispatch
EDM	Electric demand management
EED	Economic emission dispatch
EGS	Electricity-gas systems
EHs	Energy hubs
EIA	Energy information administration
EPPS	Evolutionary predator and prey strategy
EPS	Electrical power system
ER	Evidential reasoning
ESB	Energy storage battery
EUs	Energy users
FEL	Following the electric load
FTL	Following the thermal load
GB	Gas-fired boiler
GEP	Generation expansion planning
GSC	Gas supply cost
GSOMP	Group search optimizer with multiple producers
GT	Gas turbine
HC	Household air-conditioner
HE	Heat exchanger
HRSG	Heat recovery steam generator
HS	Heating system
HV	Hypervolume
IBD	Iterative benders decomposition
IER	Interval evidential reasoning
IES	Integrated energy system
IGD	Inverted generational distance
LC	Load center
LCC	Life cycle cost
LHV	Lower heating value
LMPs	Locational marginal prices
LODFs	Line outage distribution factors
LSIES	Large-scale integrated energy system
MADM	Multi-attribute decision making
MC	Monte Carlo
MCC	Multi-stage contingency-constrained co-planning
MED	Mean Euclidian distance
MGSOACC	Multi-objective group search optimizer with adaptive covariance matrix and chaotic search
MGSO-ACL	Multi-objective group search optimizer with adaptive covariance and Lévy flights
MINLP	Mixed integer nonlinear program
MODE	Multi-objective differential evolution algorithm
MOPSO	Multi-objective particle swarm optimizer

MV	Mean-variance
NG	Natural gas
NGGEP	Natural gas grid expansion planning
NPS	Number of Pareto-optimal solutions
NSGA-II	Non-dominated sorting genetic algorithm-II
OLA	Outer linearization algorithm
P2G	Power-to-gas
P2GEP	P2G plants expansion planning
PEC	Primary energy consumption
PGC	Power generation cost
PMSG	Permanent magnet synchronous generator
RC	Reciprocating chiller
SADE	Strategy adaptation based differential evolution algorithm
SC	Solar collectors
SCUC	Security constrained unit commitment
SH	Solar water heaters
SNG	Synthetic natural gas
SOO	Single objective optimization method
SP	Separation production
TDM	Thermal demand management
TEP	Transmission expansion planning
TLBO	Teaching-learning-based optimization
VFD	Variable frequency drive
WT	Wind turbine generator

Chapter 1
Introduction

Abstract Along with the popularity of distributed energy generation and hybrid energy appliances, the optimization of large-scale integrated energy systems (LSIES) combining various kinds of energy has attracted increasing attention. In this chapter, we discuss the main characteristics of LSIES. The LSIES can be depicted by the characteristics of heterogeneity, hierarchical structure, complex network topology, multimodal, hybrid variables, and complex dynamics. Additionally, conventional techniques on modeling, optimization, and decision-making of the integrated energy systems are investigated.

Keywords Large-scale integrated energy systems · Modeling methods · Multi-objective optimization · Multi-attribute decision-making

1.1 Main Characteristics of Large-Scale Integrated Energy Systems

LSIES is a typical complex system with the characteristics of heterogeneity (Jean 2015), hierarchical structure (Rodrigo et al. 2014), complex network topology (Wang et al. 2016), multimodal (Fabian et al. 2018), hybrid variables (Fiorentini et al. 2017), and complex dynamics (Jose et al. 2018). In the last decades, modeling (Wei et al. 2017), optimization (Zheng et al. 2017), and dynamic analysis (Lim et al. 2015; Jie et al. 2012) of LSIES with complicated structure were investigated quite intensively with system security and reliability fully considered (Fig. 1.1).

Currently, an active research is focused on two aspects. On one hand, LSIES is technologically more complex in structure and larger in scale. It is a typical heterogenous system combined with multiple sub-individuals: electrical power system (EPS), natural gas system (NGS), and heating system (HS) (Liu and Mancarella 2016). Each sub-individual is a hierarchical system with various network topologies (Samantha and Nathan 2018; Huang et al. 2017), and they interact with others through coupling units such as combined heat and power (CHP) (Li et al. 2017). Different characteristics of the coupling energies result in its features of heterogeneity and nonlinearity. Meanwhile, there are different components in the LSIES such as CCHP

Q.-H. Wu et al., *Large-Scale Integrated Energy Systems*, Energy Systems in Electrical Engineering, https://doi.org/10.1007/978-981-13-6943-8_1

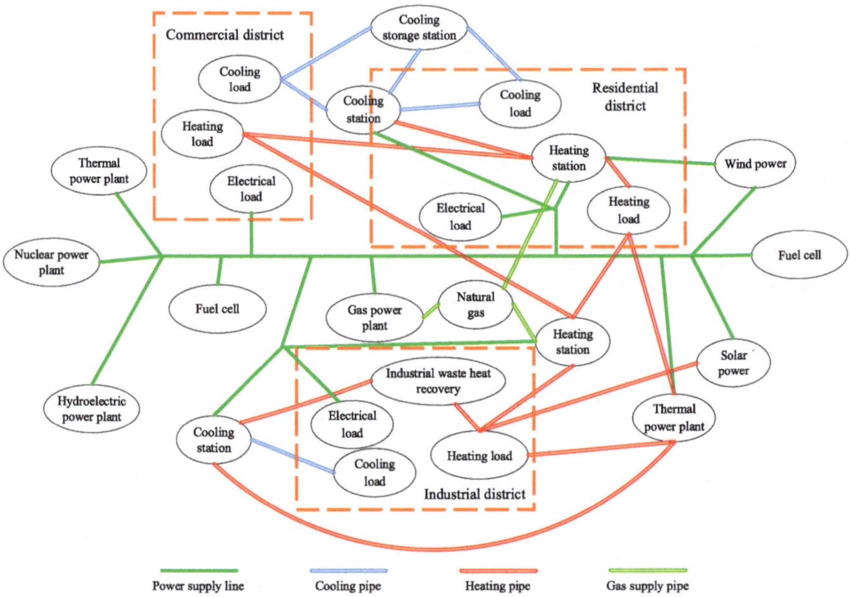

Fig. 1.1 The framework of an IES

(Li et al. 2016), heat pumps (Ommen et al. 2014), or energy storage (Hemmati 2018; Renaldi and Daniel 2017), which create lots of linkages among multi-vector energy networks. The new technologies included in the overall optimisation of energy systems may cause computational difficulty during the optimization (Amirnekooei et al. 2012; Rahmani and Amjady 2017).

Furthermore, in a competitive energy market, each individual in LSIES optimizes its own operation strategy to maximize its benefits. Due to the conflicting benefits of system individuals, only part of the information can be exchanged between each other (Nogales et al. 2003).

On the other hand, the heterogenous structure leads to hybrid variables and multimodal dynamics of LSIES. In EPS, power travels at the speed of light and usually reaches a steady state within seconds. In HS, the hydraulic process is slower, which travels at the speed of sound, taking seconds to minutes to achieve the quasi-steady state. By contrast, hot water runs even slower at the speed of mass flow rate, so it takes minutes to reach a quasi-steady state (Pan et al. 2016). These characteristics aforementioned lead to the complex dynamics and multimodal behaviour (Fabian et al. 2018) of LSIES.

1.2 Methodologies of Modeling, Optimization, and Decision-Making for Integrated Energy Systems

1.2.1 Modeling for Integrated Energy Systems

Theoretical studies on the system modeling of LSIES have occurred in recent years. Liu et al. (2016) presented an integrated electrical-hydraulic-thermal model to analysis the LSIES. Nonetheless, the optimal dispatch of heat network is not considered in order to reduce the calculation difficulty. In Liu et al. (2016), a set of relevant coupled equations which integrates electricity, heat, and gas networks was solved by Newton–Raphson method. Nevertheless, the article sets the basis for operational analysis of multi-energy systems at a district level. Once the system becomes larger, the model breaks down.

Two of the major global challenges facing the world are responding to global warming and ensuring energy security. Apart from the energy demand due to climatic conditions, rapid economic development, growing population, and higher living standards have increased the amount of energy consumption, a significant fraction of which is used for building heating. Hence, it is necessary and important for a heating system to reduce fossil fuel consumption while ensuring the quality of heating. As an important type of heating system, a district heating system is widely used to meet the increasing heat demand from the viewpoint of energy saving (Stevanovic et al. 2007; Namkhainyam et al. 2008; Gebremedhin 2012). A considerable amount of research has been conducted to reduce the fossil fuel consumption of a district heating system (Rosa and Christensen 2011; Roonprasang et al. 2008; Østergaard et al. 2010; Zheng et al. 2012; Sperling and Möller 2012; Marbe and Harvey 2006; Østergaard and Lund 2011; Ozgener 2010).

In practice, a higher resource efficiency can be achieved by combining heating and cooling in a system instead of providing these energy services separately (Xu et al. 2010; Rezaie and Rosen 2012; Chang et al. 2009). A district heating and cooling (DHC) system, which produces hot and cold fluids, and then distributes them to the residents via underground pipes, can ensure the security of supply, increase efficiency and reduce fuel costs. Furthermore, it has been proved that the integrated utilization of energy resources, including both nonrenewable energy and renewable energy, can make energy supply sustainable and reliable concurrently (Jiang et al. 2014).

Based on DHC systems, combined cooling, heating and power (CCHP) system is a kind of multi-generation energy system with cascade utilization of energy to meet multiple demands, such as electricity, thermal, and cooling energy demands. A CCHP system mainly includes the prime mover, electricity generator, heat recovery system, thermally activated equipment, and the management and control system (Cho et al. 2009; Xu and Qu 2013; Wang et al. 2014). As the rapid development of the substantial energy like photovoltaic technologies and wind turbines, wind, and solar energy are utilized more widely in the small-scale CCHP (Marano et al. 2012; Nosrat et al. 2014; Ranjbar et al. 2014). However, there is still dearth of a more refined model. On the one hand, many authors neglect the part-load performance of the generation units. As a consequence, the results are not so accurate in reality and the designed system is usually faced with a variety of problems (Yuan et al. 2015). To the author's knowledge, the difference between constant characteristic modeling and part-load characteristic modeling is significant and cannot be ignored (Zhou et al. 2013). On the other hand, further studies are still necessary to establish a relatively accurate and comprehensive model including the sustainable energy such as solar and wind energy. More attention should be paid to the influence of sustainable energy on system generation units.

Along with the popularity of the coupling units, such as DHC and CCHP, large-scale integrated energy systems (LSIES) was pushed to the fore. Researchers realized that the present energy scenario that one source's (e.g., coal, natural gas, and electricity) predominance over the others is not efficient enough for an overall mixed energy system (Quelhas et al. 2007), and a higher resource efficiency can be achieved by coupling multiple energy carriers in one system in which power can be converted arbitrarily between electrical, chemical, and thermal states instead of considering them individually (Geidl and Andersson 2005). Ana Quelhas (2007) (Quelhas and McCalley 2007) presented a multi-period generalized network flow model of the integrated energy system in the United States. The integrated system incorporates the production, storage, and transportation of coal, natural gas, and electricity in a single mathematical framework, which address the issue that the current fragmented decision-making environment in which decision-makers support strategies according to their own value system may lead to efficiency losses, and an integrated energy system and an overall optimization is needed.

LSIES is a network of multiple energy flows, such as electricity, thermal energy, cooling energy, and natural gas flows. In IES, energy can be transmitted not only in the form of electricity, but also in many forms such as heat, cold and natural gas. A wide variety of models are available for analyzing the IES.

To model the multiple energy systems, Martin Geidl first proposed a concept of Energy Hub (EH) in 2005, which exchanged power with the surrounding systems, primary energy sources, loads and other components via hybrid input and output ports (Geidl and Andersson 2007), he said that due to an increasing utilization of gas-fired and other distributed generation technologies, increased couplings between electricity, natural gas, and district heating power flow can be expected for the future (Geidl and Andersson 2005). The energy hub proposed by Martin integrates the power grid, natural gas pipeline network, district heating and cooling network, and distributed generation to provide not only electricity but also heating and cooling energy and hot water for users. Considering a system composed of several hubs, the model can be split into two parts: hub power flow and network power flow. Each model can be stated as a matrix, and based on this, Martin conducted more in-depth research in the following articles (Geidl and Andersson 2005, 2007; Geidl et al. 2007).

Based on EH, a large amount of research has been done over recent years. But most research (Parisio et al. 2011; Shahmohammadi et al. 2011; Zhang et al. 2015) in this area paid more attention to the hub internal without considering the network outside, and modeling of CCHP microgrid is a good example. Research on modeling of CCHP concentrates more on type selection of prime movers or chillers without considering energy flow of the whole network (Gu et al. 2014).

Modeling of LSIES considering network energy flow is a hot topic in the area of planning and operation of LSIES. Scala (2014) addressed the problem of optimal energy flow management in multi-carrier energy networks in the presence of integrated energy hubs. In his research, a gas pipeline was assumed to be the same as an electrical network, and the bus voltage and the natural gas pressures in all gas pipelines are simulated under hourly changed energy price and thermal load, the results showed that compared to the voltage magnitudes and angles, the natural gas pressures were always in a relatively stable state. Moein (2014) proposed a model of coupling gas and electricity network and conclude that with energy hubs, the local marginal costs of electricity and gas were decreased. It's noteworthy that these researchers both concentrated on the relatively stable gas pipeline network instead of the cooling and heating pipeline network which is not so stable and may cause significant energy loss.

As mentioned above, researchers preferred integrated models (Liu et al. 2016; Shabanpour-Haghighi and Seifi 2015, 2016) of natural gas, electricity, and heating and cooling energy. The modeling is not accurate enough and difficult to be applied to large-scale systems. An accurate integrated model is usually large, non-convex, too time-consuming to solve, or even not convergent. In order to deal with this problem, lots of simplifications need to be done to the integrated model. Therefore, some important characteristics of the system may be lost. Besides, since the models are established and solved in an integrated manner, it is difficult to modify the coupled model to increase or decrease components.

In an integrated model, assumptions are made that the participants of the system completely obey the arrangement of the whole system without considering their own interests, which is the traditional operating paradigm developed for regulated industries. Nevertheless, recent move to an open up energy market where system individuals can make decisions on their own has touched off the requests for operational independence and information privacy. It is obvious that traditional centralized optimization methods which need all information about the system are not applicable for solving optimal operation problem of LSIES.

Traditional modeling methods for complex systems are mainly divided into two types, one is linear modeling method, and the other is nonlinear modeling method. For linear modeling method, state-space equations are widely used and simplifications are done to complex systems. In this way, many details on important characteristics of complex systems are lost. Additionally, hundreds of equations of a complicated system would make the model hard to solve. For nonlinear modeling method, developers model one same system in various ways. The accuracy, the boundary constraints, and the order of equations may vary a lot from one model to another. This results to the difficulty of comparison with disparate solutions drawn from one system. Therefore, a more common and standard model which can distinctly represent a complex system is essential.

1.2.2 Optimization for Integrated Energy Systems

The optimization methods of LSIES is another focus of attention. Shabanpour-Haghighi (2015, 2016) utilized teaching-learning based optimization method to solve energy flow of electricity, gas, and heat combined networks. The objective function was to minimize the total operating costs of the system, which was composed of the costs of natural gas, coal, biomass, and so on. A set of nonlinear equations based on the nodal power balance of each network was solved by Newton–Raphson method. In Liu et al. (2016), a multi-temporal simulation model was proposed to model a multi-vector district energy system which integrates electricity, heat and gas networks in an integrated manner. Then the model was applied to a real case study of University of Manchester to validate the efficiency of the optimization method. Liu and Mancarella (2016) emphasized that since CCHP, heat pumps, and electric boilers create lots of linkages among multi-vector energy networks, an integrated electrical-hydraulic-thermal method will play an important role in the analysis of IES. Jing et al. (2014) presented a comprehensive model of integrated energy based district heating and cooling system, and an optimal operating strategy was investigated to optimize the daily running cost of the whole system. Zheng et al. (2015) developed a multi-objective group search optimizer to optimize the power dispatch of a LSIES with the selected objectives compromising the competing benefits of both the power grid and the district heating and cooling units.

The optimization problems of LSIES may have one single-objective function which is typically related to economic performances. However, the energy supply reliability and CO_2 emission reduction are getting more and more important in the analysis of LSIES's performance. Decreasing one of the different objectives would increase the other one and vice versa. Therefore, a multi-objective optimization could be utilized to obtain the global optima of the problems based on the priority of each objective (Shabanpour-Haghighi and Seifi 2015).

A many-objective optimization problems (MaOPs) is a special kind of multi-objective optimization problems (MOPs) with more than three objectives. With the multiple conflicting demands faced by the industry today for superior quality, low cost, higher safety, and so on, competitive edge could only be established by designing the products and processes that account for as many performance objectives as possible. It implies that many objectives need to be simultaneously dealt with, in an optimization problem. However, it is hard to obtain the entire solution set of a many-objective optimization problems (MaOPs) by multi-objective optimization algorithms (such as NSGA-, GSOMP) because of the difficulties brought by the curse of dimensionality (Wang and Yao 2016) (Fig. 1.2).

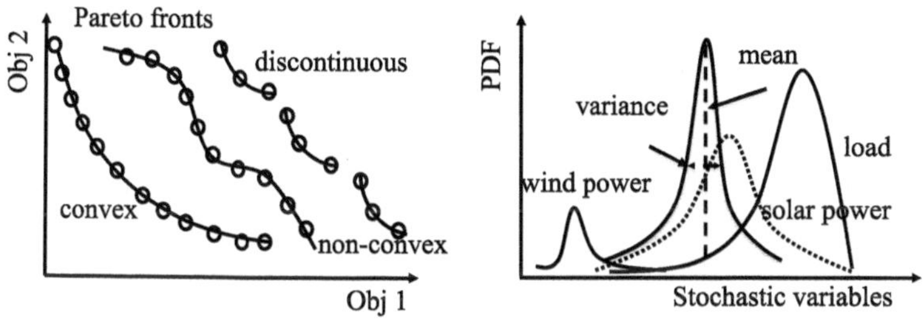

Fig. 1.2 The relationships of objective functions in the multi-objective optimization

1.2.3 Decision-Making for Integrated Energy Systems

The major impediments in handling large number of objectives relate to stagnation of search process (Deb and Saxena 2005), inefficiency of selection operators, high computational cost, and difficulty in visualization of objective space (Saxena et al. 2013).

(1) Inefficiency of selection operators: with the increase in M, multi-objective problem may well have a high-dimensional Pareto-set with complicated shapes, this makes the most existing MOEAs and MGSO ineffective, whose primary selection based on Pareto-dominance.

(2) High computational cost: if a continuous multi-objective optimization problem (with M objectives) meets the regularity property, the dimension of its POF can be $M - 1$. Therefore, the number of points needed to approximate the whole POF increase exponentially with M. The same phenomenon can be observed in discrete problems.

(3) Difficulty in visualization of a POF for problems with $M \geq 4$: finding a higher dimensional Pareto-optimal surface is one important matter, but visualizing it for a proper decision-making is another equally important matter.

There are more and more related researches that have been put forward to overcome the difficulty of MaOPs, they can be roughly divided into the following three classes:

(1) Scalarization technique: all objectives are converted into a single composite objective by the weighed sum approach. With different vectors repeating the scalarization process, there will be a number of Pareto-optimal solutions, then decision-making can be implemented to complete the optimization. However, each Pareto-optimal solution is independent to each other, thereby losing the parallel search ability often desired in solving complex optimization problems.

(2) Involvement of decision-makers (DM): for a large number of objectives, it can involve a decision-maker right in the beginning of the optimization process and instead of finding the optimal solutions corresponding to a specific weight-vector or varepsilon-vector, although this beats the dimensionality problem described earlier by not finding points on the complete high-dimensional Pareto-optimal frontier and also providing the decision-maker with a set of solutions in a region of interest to

her or him, but this approach may only be a portion of the true Pareto-optimal front, thereby reducing the number and dimensionality of target solutions, arriving at a biased distribution of Pareto-optimal solutions (Deb and Saxena 2005).

(3) Objective reduction: because of considering the importance of complete Pareto-optimal front, this approach is more suitable for solving the MaOPs with redundant objectives. The greatest difficulty of MaOPs is the curse of dimensionality, that is to say if N points are needed for adequately representing a one-dimensional Pareto-optimal front, N_m points will be necessary to represent an M-dimensional Pareto-optimal front. A lower dimensional Pareto-optimal front can be possessed by eliminating objectives that are redundant.

In the absence of any other information, none of the generated Pareto-optimal solutions can be said to be better than the other. Thus, preference information is in demand to determine the ROI, and different types of preference information asked from the DM include marginal rates of substitution, surrogate values for trade-offs, classification of objective functions and reference points (Ruiz et al. 2015; Miettinen and MaKela 2002). A reference point consists of aspiration levels, that is, objective function values that are desirable or acceptable for the DM. However, there are additional cases of more than one decision maker with conflicting interest and different importance level from a hierarchical viewpoint. We need to employ a negotiation support system (Bechikh et al. 2013) to aggregate the DMs conflicting preferences.

It must be noted that the grid for the objective functions in the AUGMECON2 does not need to be constant throughout the whole process, but can be adjusted. The number of efficient solutions generated AUGMECON2 is determined by the length of equal interval, the longer the length of equal interval, the less the number of efficient solutions. That is, the number of effective solutions is certain and artificially controlled. As such, the interactive methods and the AUGMECON2 is a good match to assist the DM to identify the most preferred efficient solution after a number of iterations. In this way, the DM may start with a coarse grid (longish equal interval) in the early iterations to optimize MOPs or MaOPs, in order to cover quickly the whole Pareto front. Subsequently, in the late iterations, when the DM is making fine tuning, they can accordingly use a denser grid for investigating more thoroughly the ROI. In the following section, we assume that the reference point below is the reference point after negotiation. A reference point is given by $\mathbf{q} = (q_1, q_2, \ldots, q_k)^{\mathcal{T}}$, where q_i is an aspiration value for the objective function f_i provided by the DM, for all $i = 1, \ldots, k$. Usually, a reference point is achievable if there is a feasible solution whose objective values simultaneously achieve or improve the corresponding reference levels; otherwise, the reference point is unachievable.

1.3 Contents of This Book

This book is organized as follows.

Chapter 2 introduces the modeling of large-scale integrated energy systems, which consist of district heating and cooling systems, combined cooling heating and Power systems, energy hubs, and individual-based model for the LSIES. First the DHCs are presented, including DHC station, water supply network, heating load, and cooling load. In the modeling of CCHP, the wind turbine generator, gas turbine, and heat recovery generator and load center are described. Moreover, the concept of energy hubs is introduced, including gas turbine, energy storage battery, heat exchanger and so on. Finally, an individual-based model is defined for LSIES along with several IBMs for test IES.

Chapter 3 focuses on the multi-objective optimization algorithms dealing with the multi-objective optimization problems. First, a multi-objective group search optimizer with adaptive covariance and Lévy flights is presented with three groups applying different search strategies. Moreover, this chapter presents a novel multi-objective evolutionary predator and prey strategy (EPPS) employing novel search mechanisms to guarantee the global and local searching abilities. Comparisons between the EPPS and the state-of-the-art algorithms are summarized in this chapter.

Chapter 4 concentrates on the introduction of multi-attribute decision-making support system, including an improved entropy weight method, evidential reasoning, and interval evidential reasoning. Furthermore, the relationship between different objectives are analyzed using a framework of reduction for nonlinear many objectives. The performance test is carried out and comparisons between these multi-attribute decision-making analysis methods are discussed.

Chapter 5 deals with the planning problems of the LSIES considering the optimal unit sizing for the LSIES and the multistage contingency-constrained co-planning for the LSIES. In the unit sizing problem, the multi-objective interval optimization model is first presented. Then the corresponding optimization and decision-making methods are raised to solve the planning problem. In the multistage contingency-constrained (MCC) co-planning problem, the MCC co-planning model is developed with the consideration of contingency constraints. Afterwards, the solution methodology for the co-planning problem is introduced along with case studies in two different simulation test systems.

Chapter 6 is concerned with the optimal operation of LSIES considering multiple operating interests and various operating scenarios. Firstly, the optimal operation of LSIES with distributed DHCs embedded is illustrated. The operating models of electricity network and gas network are introduced, and the coordinated scheduling strategy for the integrated electricity and gas networks is developed. Then an energy trading game for LSIES considering conflict benefits is studied in this chapter. The energy trading game of DESs and EUs integrated into the LSIES is introduced, and simulation studies using a hybrid multi-objective optimization and game-theoretic approach are conducted on a test IES.

1.4 Summary

In this chapter, the background of LSIES and the modeling, optimization and decision-making for the LSIES are introduced first. The main characteristics of LSIES has been presented and compared with the traditional power systems. Then the existing methodologies of modeling, optimization, and decision-making for LSIES are summarized. Finally, the book outline is provided to give a clear view of the entire contents.

References

Amirnekooei K, Ardehali MM, Sadri A (2012) Integrated resource planning for iran: Development of reference energy system, forecast, and long-term energy-environment plan. Energy 46(1):374–385

Bechikh S, Said LB, Ghdira K (2013) Group preference-based evolutionary multi-objective optimization with non-equally important decision makers: application to the portfolio selection problem 5:1

Chang WS, Wang CC, Shieh CC (2009) Design and performance of a solar-powered heating and cooling system using silica gel/water adsorption chiller. Appl Therm Eng 29(10):2100–2105

Cho H, Mago PJ, Luck R, Chamra LM (2009) Evaluation of cchp systems performance based on operational cost, primary energy consumption, and carbon dioxide emission by utilizing an optimal operation scheme. Appl Energy 86(12):2540–2549

Deb K, Saxena DK (2005) On finding pareto-optimal solutions through dimensionality reduction for certain large-dimensional multi-objective optimization problems. Kangal report 2005011

Fabian S, Balthasar B, Hendrik K (2018) Towards integrated multi-modal municipal energy systems: an actor-oriented optimization approach. Appl Energy 228:2009–2023

Fiorentini M, Wall J, Ma Z, Braslavsky JH, Cooper P (2017) Hybrid model predictive control of a residential hvac system with on-site thermal energy generation and storage. Appl Energy 187:465–479

Gebremedhin A (2012) Introducing district heating in a Norwegian town–potential for reduced local and global emissions. Appl Energy 95:300–304

Geidl M, Andersson G (2007) Optimal power flow of multiple energy carriers. IEEE Trans Power Syst 22(1):145–155

Geidl M, Koeppel G, Favre-Perrod P, Klockl B, Andersson G, Frohlich K (2007) Energy hubs for the future. IEEE Power Energy Mag 5(1):24–30

Geidl M, Andersson G (2005) A modeling and optimization approach for multiple energy carrier power flow. In: 2005 IEEE Russia Power Tech, pp. 1–7

Geidl M, Andersson G (2005) Operational and topological optimization of multi-carrier energy systems. In: 2005 International Conference on Future Power Systems, p 6. IEEE

Geidl M, Andersson G (2007) Optimal coupling of energy infrastructures. In: 2007 IEEE Lausanne Power Tech, pp. 1398–1403

Gu W, Wu Z, Bo R, Liu W, Zhou G, Chen W, Wu Z (2014) Modeling, planning and optimal energy management of combined cooling, heating and power microgrid: A review. Int J Electr Power Energy Syst 54:26–37

Hemmati R (2018) Optimal design and operation of energy storage systems and generators in the network installed with wind turbines considering practical characteristics of storage units as design variable. J Clean Prod 185:680–693

Huang JB, Li Z, Wu QH (2017) Coordinated dispatch of electric power and district heating networks: A decentralized solution using optimality condition decomposition. Appl Energy 206:1508–1522

Jean MC, Nadia M (2015) Integrating household behavior and heterogeneity into the times-households model. Appl Energy 139:56–67

Jiang XS, Jing ZX, Li YZ, Wu QH, Tang WH (2014) Modeling and operation optimization of an integrated energy based direct district water-heating system. Energy 64:375–388

Jie PF, Tian Z, Yuan SS, Zhu N (2012) Modeling the dynamic characteristics of a district heating network. Energy 39(1):126–134

Jing ZX, Jiang XS, Wu QH, Tang WH, Hua B (2014) Modelling and optimal operation of a small-scale integrated energy based district heating and cooling system. Energy 73:399–415

Jose DM, Carlos JF, Fabiola A (2018) Simulation of demand growth scenarios in the colombian electricity market: an integration of system dynamics and dynamic systems. Appl Energy 216:504–520

Li M, Mu H, Li N, Ma B (2016) Optimal design and operation strategy for integrated evaluation of cchp (combined cooling heating and power) system. Energy 99:202–220

Li GQ, Zhang RF, Jiang T (2017) Optimal dispatch strategy for integrated energy systems with cchp and wind power. Appl Energy 192:408–419

Lim S, Park S, Chung H, Kim M (2015) Dynamic modeling of building heat network system using simulink. Appl Therm Eng 84:375–389

Liu XZ, Mancarella P (2016) Modelling, assessment and sankey diagrams of integrated electricity-heat-gas networks in multi-vector district energy systems. Appl Energy 167:336–352

Liu XZ, Wu JZ, Jenkins N, Bagdanavicius A (2016) Combined analysis of electricity and heat networks. Appl Energy 162:1238–1250

Marano V, Rizzo G, Tiano FA (2012) Application of dynamic programming to the optimal management of a hybrid power plant with wind turbines, photovoltaic panels and compressed air energy storage. Appl Energy 97:849–859

Marbe Å, Harvey S (2006) Opportunities for integration of biofuel gasifiers in natural-gas combined heat-and-power plants in district-heating systems. Appl Energy 83(7):723–748

Miettinen K, MaKela MM (2002) On scalarizing functions in multiobjective optimization. OR Spectr 24(2):193–213

Moeini-Aghtaie M, Abbaspour A, Fotuhi-Firuzabad M, Hajipour E (2014) A decomposed solution to multiple-energy carriers optimal power flow. IEEE Trans Power Syst 29(2):707–716

Namkhainyam B, Enkhjargal K, Purevjal O (2008) Improvement energy efficiency of district heating systems. In: Proceedings of 3rd International Forum on Strategic Technologies, pp. 545–548. Novosibirsk-Tomsk, Russia

Nogales FJ, Prieto FJ, Conejo AJ (2003) A decomposition methodology applied to the multi-area optimal power flow problem. Ann Oper Res 120(1):99–116

Nosrat AH, Swan LG, Pearce JM (2014) Simulations of greenhouse gas emission reductions from low-cost hybrid solar photovoltaic and cogeneration systems for new communities. Sustain Energy Technol Assessments 8:34–41

Ommen T, Markussen WB, Elmegaard B (2014) Heat pumps in combined heat and power systems. Energy 76:989–1000

Østergaard PA, Lund H (2011) A renewable energy system in Frederikshavn using low-temperature geothermal energy for district heating. Appl Energy 88(2):479–487

Østergaard PA, Mathiesen BV, Möller B, Lund H (2010) A renewable energy scenario for Aalborg Municipality based on low-temperature geothermal heat, wind power and biomass. Energy 35(12):4892–4901

Ozgener O (2010) Use of solar assisted geothermal heat pump and small wind turbine systems for heating agricultural and residential buildings. Energy 35(1):262–268

Pan ZG, Guo QL, Sun HB (2016) Interactions of district electricity and heating systems considering time-scale characteristics based on quasi-steady multi-energy flow. Appl Energy 167:230–243

Parisio A, Del Vecchio C, Velotto G (2011) Robust optimization of operations in energy hub. In: 2011 50th IEEE Conference on Decision and Control and European Control Conference (CDC-ECC), pp. 4943–4948 (2011)

Quelhas A, Gil E, McCalley JD, Ryan SM (2007) A multiperiod generalized network flow model of the US integrated energy system: Model description. IEEE Trans Power Syst 22(2), 829–836 (2007)

Quelhas A, McCalley J (2007) A multiperiod generalized network flow model of the US integrated energy system: simulation results. IEEE Trans Power Syst 22(2), 837–844 (2007)

Rahmani S, Amjady N (2017) A new optimal power flow approach for wind energy integrated power systems. Energy 134:349–359

Ranjbar MR, Mohammadian M, Esmaili S (2014) Economic analysis of hybrid system consists of fuel cell and wind based CHP system for supplying grid-parallel residential load. Energy Build 68, Part A(0), 476–487
Renaldi R, Daniel F (2017) Multiple time grids in operational optimisation of energy systems with short- and long-term thermal energy storage. Energy 133:784–795
Rezaie B, Rosen MA (2012) District heating and cooling: Review of technology and potential enhancements. Appl Energy 93:2–10
Rodrigo M, Martin H, Yan F, Enrico Z (2014) Self-adaptable hierarchical clustering analysis and differential evolution for optimal integration of renewable distributed generation. Appl Energy 133:388–402
Roonprasang N, Namprakai P, Pratinthong N (2008) Experimental studies of a new solar water heater system using a solar water pump. Energy 33(4):639–646
Rosa AD, Christensen JE (2011) Low-energy district heating in energy-efficient building areas. Energy 36(12):6890–6899
Ruiz AB, Saborido R, Luque M (2015) A preference-based evolutionary algorithm for multiobjective optimization: the weighting achievement scalarizing function genetic algorithm. J Glob Optim 62(1):101–129
Samantha AJ, Nathan GJ (2018) Scalable multi-agent microgrid negotiations for a transactive energy market. Appl Energy 229:715–727
Saxena DK, Duro JA, Tiwari A, Deb K, Zhang Q (2013) Objective reduction in many-objective optimization: linear and nonlinear algorithms. IEEE Trans Evol Comput 17(1):77–99
Scala ML, Vaccaro A, Zobaa AF (2014) A goal programming methodology for multiobjective optimization of distributed energy hubs operation. Appl Therm Eng 71(2):658–666
Shabanpour-Haghighi A, Seifi AR (2015) Simultaneous integrated optimal energy flow of electricity, gas, and heat. Energy Convers Manag 101:579–591
Shabanpour-Haghighi A, Seifi AR (2015) Multi-objective operation management of a multi-carrier energy system. Energy 88:430–442
Shabanpour-Haghighi A, Seifi AR (2016) An integrated steady-state operation assessment of electrical, natural gas, and district heating networks. IEEE Trans Power Syst 31(5):3636–3647
Shahmohammadi A, Dalvand M, Ghazizadeh M, Salemnia A (2011) Energy hubs' structural and operational linear optimization with energy storage elements. In: 2011 2nd International Conference on Electric Power and Energy Conversion Systems (EPECS), pp. 1–6
Sperling K, Möller B (2012) End-use energy savings and district heating expansion in a local renewable energy system-a short-term perspective. Appl Energy 92:831–842
Stevanovic VD, Prica S, Maslovaric B, Zivkovic B, Srdjan N (2007) Efficient numerical method for district heating system hydraulics. Energy Convers Manag 48(5):1536–1543
Wang JL, Wu JY, Zheng CY (2014) Simulation and evaluation of a CCHP system with exhaust gas deep-recovery and thermoelectric generator. Energy Convers Manag 86:992–1000
Wang MG, Tian LX, Du RJ (2016) Research on the interaction patterns among the global crude oil import dependency countries: a complex network approach. Appl Energy 180:779–791
Wang H, Yao X (2016) Objective reduction based on nonlinear correlation information entropy. Soft Comput 20(6):2393–2407
Wei F, Jing ZX, Wu PZ, Wu QH (2017) A stackelberg game approach for multiple energies trading in integrated energy systems. Appl Energy 200:315–329
Xu DH, Qu M (2013) Energy, environmental, and economic evaluation of a CCHP system for a data center based on operational data. Energy Build 67:176–186
Xu JZ, Sui J, Li BY, Yang ML (2010) Research, development and the prospect of combined cooling, heating, and power systems. Energy 35(11):4361–4367
Yuan ZX, Jing ZX, Hu RX, Wu QH (2015) Operation optimization of cchp-type microgrid considering units' part-load characteristics. In: Smart Grid Technologies-Asia (ISGT ASIA), 2015 IEEE Innovative, pp. 1–7
Zhang XP, Shahidehpour M, Alabdulwahab A, Abusorrah A (2015) Optimal expansion planning of energy hub with multiple energy infrastructures. IEEE Trans Smart Grid 6(5):2302–2311

Zheng GZ, Li F, Tian Z, Zhu N, Li QR, Zhu H (2012) Operation strategy analysis of a geothermal step utilization heating system. Energy 44(1):458–468

Zheng JH, Chen JJ, Wu QH, Jing ZX (2015) Multi-objective optimization and decision making for power dispatch of a large-scale integrated energy system with distributed DHCs embedded. Appl Energy 154:369–379

Zheng JH, Wu QH, Jing ZX (2017) Coordinated scheduling strategy to optimize conflicting benefits for daily operation of integrated electricity and gas networks. Appl energy 192:370–381

Zhou Z, Liu P, Li Z, Pistikopoulos EN, Georgiadis MC (2013) Impacts of equipment off-design characteristics on the optimal design and operation of combined cooling, heating and power systems. Comput Chem Eng 48(0), 40–47

Chapter 2
Modeling of Large-Scale Integrated Energy Systems

Abstract Large-scale integrated energy systems are networks of various energy flows, such as electricity, thermal energy, cooling energy, and natural gas flows. In a LSIES, energy can be transmitted not only in the form of electricity, but also in many other forms such as thermal energy and natural gas. A wide variety of models have drawn much attention, such as district heating and cooling systems, combined cooling heating and power systems, and energy hubs. Each of these models is described in detail in the following sections. This chapter presents the models of subsystems of IES, such as district heating and cooling systems, combined cooling heating and power systems. Moreover, we propose an individual-based model (IBM) for modeling LSIES. An individual is a basic unit consisting of a quintuple of input, knowledge, state, function, and output sets. It can make decisions independently according to accurate evolutionary mechanisms described by the function set. Additionally, the individuals interact with others through input and output sets in a unified form. In this way, a complex system can be decoupled into several independent individuals whose internal characteristics are fully specified and hidden from the external environment.

Keywords Individual-based model · District heating and cooling systems · Combined cooling heating and power systems · Energy hubs · Large-scale integrated energy systems

2.1 District Heating and Cooling Systems

For the sake of simplicity, we have made the following assumptions for this model:

(1) The integrated energy-based DHC system is proposed for the residential buildings in hot-summer and cold-winter periods. The system operates in the heating mode in winter and conversely cooling mode in summer.
(2) The proposed system is a small-scale system, using direct heating and direct cooling because of the small heating and cooling areas.
(3) The site selection of a district heating and cooling station and the laying of pipelines fully take into account the fairness of energy supply, and the supply water temperature for each building is assumed to be the same.

Q.-H. Wu et al., *Large-Scale Integrated Energy Systems*, Energy Systems in Electrical Engineering, https://doi.org/10.1007/978-981-13-6943-8_2

(4) The heat or cold losses of pipelines caused by water losses are neglected, here we just consider heat and cold losses caused by thermal convection.
(5) Each building is regarded as a whole from the viewpoint of a network. The dry bulb temperature or relative humidity inside each building is set as an average value.

2.1.1 Description of the DHC System

Figure 2.1 illustrates a two-pipe direct district heating and cooling system located in a residential area. The system includes a district heating and cooling station, a water supply network, and terminal units. The district heating and cooling station consist of two off-grid wind generators (Permanent Magnet Synchronous Generator, PMSG), three absorption chillers, two reciprocating chillers, and heat producers including a gas-fired hot water boiler, an electric water boiler, and a flat-plate solar water heater. The water network, including water pipes, hot water pumps (P1, P3, P5, ..., P2n+1), and chilled water pumps (P2, P4, P6, ..., P2n+2), are employed to transport hot water or cold water from the station side to terminal units in resident buildings. Heating radiators and fan coil units are used as the terminal units for heating and cooling of buildings, respectively. At the station side, valves (V3, V4, V5, V6, V7, V8, V9) are closed and valves (V1, V2) are open during a heating season. During this season, the chillers are not put into activated operation and the system just operates in the heating mode. Conversely, valves (V3, V4, V5, V6, V7, V8, V9) are open and valves (V1, V2) are closed during the cooling season. The absorption chillers and the reciprocating chillers use thermal energy and electric energy as the driving force, respectively, to supply chilled water for cooling in a coordinative manner. While the operation mode of the system is changed, the set of circulating pumps being used will be switched to the corresponding condition, as well as terminal units. It should be noted that the circuit breaker (QF1) is switched off and the circuit breaker (QF2) is turned on in a cooling season. This is because electricity-driven compression cooling is much more energy efficient than heat-driven absorption cooling while electrical power is utilized as an energy source.

2.1.2 District Heating and Cooling Station

2.1.2.1 Solar Water Heater

As we know, solar energy is uncertain and intermittent. In consideration of stability and reliability, the water heater is used for water preheating as shown in Fig. 2.1. A steady-state model of the water heater can be expressed by (Zhai et al. 2007)

$$P_{\mathrm{H3}} = \eta_{\mathrm{c}} N_{\mathrm{col}} A_{\mathrm{c}} H_{\mathrm{T}} = M_{\mathrm{O}} c_{\mathrm{w}} \left(T_{\mathrm{o3}} - T_{\mathrm{i3}}\right) \tag{2.1}$$

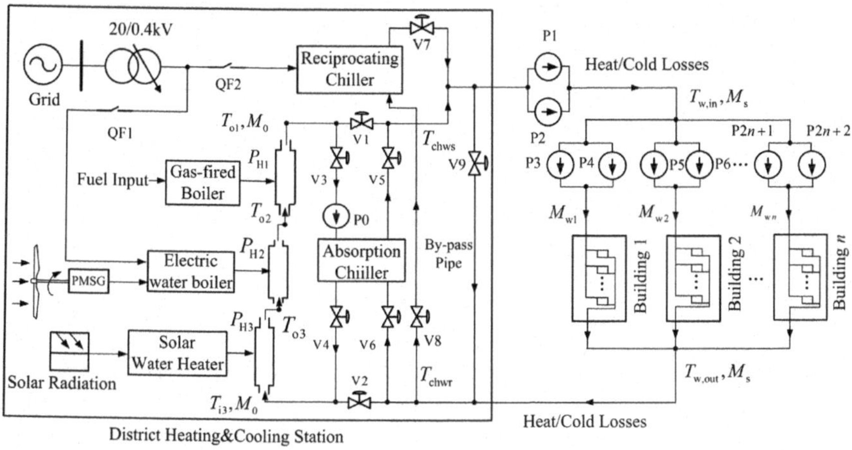

Fig. 2.1 Schematic diagram of a DHC system

where P_{H3} is the heat production of the heater, η_c is the efficiency of the solar collector, N_{col} is the number of solar collectors present at the site, A_c is the area of a solar collector, H_T is the total solar flux incident on the tilted collector, M_0 is the mass flow of water through the tank, and T_{i3} and T_{o3} are the inlet temperature and the outlet temperature of the tank, respectively.

2.1.2.2 Off-grid Wind Power System

An off-grid wind turbine generator directly connected to resistive elements is used for water heating. The detailed introduction to its model and control strategies can be referred to Jiang et al. (2014). For each wind speed, the resistance of heating elements can be adjusted to a desired value according to a reference electrical rotating speed so that the generator can operate at an expected output.

2.1.2.3 Electric Water Boiler

The energy supplied to the electric water boiler comes from the wind turbine generators and the power grid. The steady-state model of the electric boiler can be described as (Dolan et al. 1996; Paull et al. 2009):

$$P_{H2} = \eta_2 \left(P_{pur} + P_{WG}\right) = M_0 c_w \left(T_{o2} - T_{i2}\right) \tag{2.2}$$

where P_{H2} and η_2 are the heat production and the operation efficiency of the electric water boiler, respectively, P_{pur} is the electric power purchased from the power grid,

M_0 is the mass flow of water through the tank, c_w is the specific heat of water, and T_{i2} and T_{o2} are the inlet temperature and the outlet temperature of the tank, respectively.

2.1.2.4 Gas-fired Water Boiler

A low temperature central gas-fired boiler investigated in Spreitzer et al. (2002) is used for water heating. Under steady-state conditions, the model can be expressed as

$$P_{H1} = \eta_1 q_g B_g = M_0 c_w (T_{o1} - T_{i1}) \tag{2.3}$$

where P_{H1} and η_1 are the heat production and the operation efficiency of the gas-fired water boiler, respectively, q_g is the calorific value of natural gas, B_g is the natural gas consumption rate, M_0 is the mass flow of water through the water compartment, and T_{i1} and T_{o1} are the inlet temperature and the outlet temperature of the water compartment, respectively.

2.1.2.5 Reciprocating Chiller

Reciprocating chillers are used in a wide range of commercial and industrial applications, and represent a substantial fraction of installed cooling systems. In this research, we consider chiller operation at steady state only, namely, turnon and turnoff transients are neglected. For a reciprocating chiller, its cooling rate can be expressed as

$$Q_{c1} = \mathrm{COP}_1 \cdot P_{pur} = M_{c1} c_w \left(T^{in}_{evap\,1} - T^{out}_{evap\,1} \right) \tag{2.4}$$

where COP_1 is the coefficient of performance of a reciprocating chiller, which is defined as the cooling rate Q_{c1} divided by the electric power input P_{pur}, M_{c1} is the mass flow rate of chilled water through the chiller, and $T^{in}_{evap\,1}$ and $T^{out}_{evap\,1}$ are the evaporator inlet temperature and outlet temperature of a reciprocating chiller, respectively.

A simple model for predicting the performance of reciprocating chillers was utilized (Gordon and Ng 1994). The model predicts specific functional dependences on the key system variables, with three adjustable parameters that characterize chiller irreversibility, which is described as follows:

$$\frac{1}{\mathrm{COP}_1} = -1 + \frac{T^{in}_{cond\,1}}{T^{out}_{evap\,1}} + \frac{-A_0 + A_1 T^{in}_{cond\,1} - A_2 \left(T^{in}_{cond\,1} / T^{out}_{evap\,1} \right)}{Q_{evap_1}} \tag{2.5}$$

where $T^{in}_{cond\,1}$ is the condenser inlet temperature of a reciprocating chiller, Q_{evap_1} is the heat transfer to the evaporator, $Q_{evap_1} = Q_{c1}$, and the constants A_0, A_1, and A_2 characterize the irreversibility (entropy generation) of a particular chiller.

2.1.2.6 Absorption Chiller

Absorption chillers have become attractive as environmental concerns are accounted for in chiller sector, in particular when either natural gas for firing the chiller is abundant, or when low-grade thermal energy is available. Taking into account steady state only, the cooling rate can be obtained as

$$Q_{c2} = \text{COP}_2 \cdot Q_{\text{gen}} = M_{c2} c_{\text{w}} \left(T_{\text{evap}2}^{\text{in}} - T_{\text{evap}2}^{\text{out}} \right) \tag{2.6}$$

where COP_2 is the coefficient of performance of an absorption chiller, which is defined as the cooling rate Q_{c2} divided by the input thermal power to the generator Q_{gen}, M_{c2} is the mass flow rate of chilled water through the chiller, and $T_{\text{evap}2}^{\text{in}}$ and $T_{\text{evap}2}^{\text{out}}$ are the evaporator inlet temperature and outlet temperature of an absorption chiller, respectively.

A simple two-adjustable parameter model for predicting the performance presented was utilized (Gordon and Ng 1995). The approximate formula for COP_2 is given by

$$\begin{aligned} \frac{1}{\text{COP}_2} = & \left(\frac{T_{\text{cond}2}^{\text{in}} - T_{\text{evap}2}^{\text{out}}}{T_{\text{evap}2}^{\text{out}}} \right) \left(\frac{T_{\text{gen}}^{\text{in}}}{T_{\text{gen}}^{\text{in}} - T_{\text{cond}2}^{\text{in}}} \right) + \\ & \frac{1}{Q_{\text{evap}2}} \left(\frac{T_{\text{gen}}^{\text{in}}}{T_{\text{gen}}^{\text{in}} - T_{\text{cond}2}^{\text{in}}} \right) \left(A_3 - A_4 \frac{T_{\text{cond}2}^{\text{in}}}{T_{\text{gen}}^{\text{in}}} \right) \end{aligned} \tag{2.7}$$

where $T_{\text{cond}2}^{\text{in}}$ and $T_{\text{gen}}^{\text{in}}$ are the condenser inlet temperature and the generator inlet temperature of an absorption chiller, respectively, and the constants A_3 and A_4 characterize the irreversibility (entropy generation) of a particular chiller.

2.1.3 *Water Supply Network*

2.1.3.1 Pressure Drop Model

To design the control strategy for circulating pumps, a water network pressure drop model that characterizes the pressure drop on each individual component in the system is developed. Figure 2.2 illustrates the simple structure of water network pressure drop for the system. Variable speed circulating pumps are employed to overcome the friction of a piping system and ensure the hydraulic balance of a heating system. Pump curves and piping head loss curve are shown in Fig. 2.3. ω_{p} represents the rotational speed of a pump. As observed from Fig. 2.3, the operation point of a pump is the intersection of both the piping head loss curve and the pump curve (Ma and Wang 2009).

The pumps installed in the main supply pipeline is used to overcome the piping pressure drop, which includes the pressure drop across the heat units or chillers,

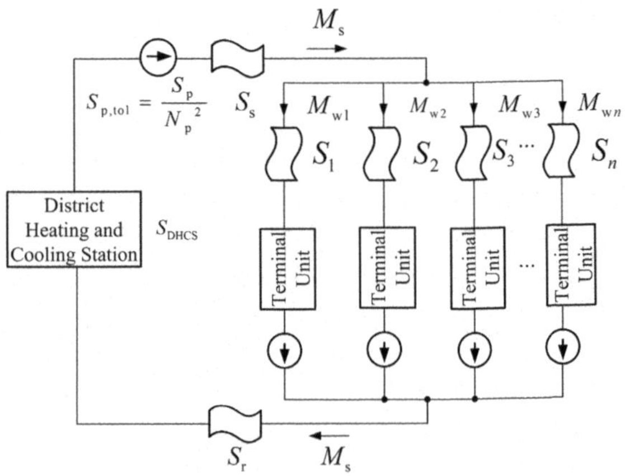

Fig. 2.2 Structure of the water network pressure drop model

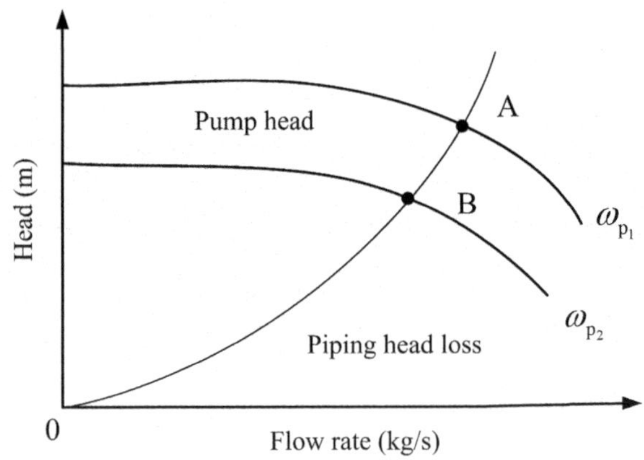

Fig. 2.3 Characteristics of pump head and piping head loss

the fittings around pumps and the main supply and return pipeline. This can be mathematically described as

$$H_0 = \left(S_{DHCS} + S_{p,tol} + S_s + S_r\right) M_s^2 \tag{2.8}$$

where H_0 is the head of pump installed in the main supply pipeline, M_s is the water flow rate of main pipelines, and S_{DHCS}, $S_{p,tol}$, S_s and S_r are the flow resistance of operating heat units or chillers, fittings around pumps, supply pipeline and return pipeline, respectively.

The pump installed in the jth branch is used to overcome the friction of the jth branch pipeline, which can be expressed as

$$H_j = S_j M_{wj}^2 \tag{2.9}$$

where H_j is the head of pump installed in the jth branch, and S_j and $M_{\mathrm{w}j}$ are the flow resistance and the flow rate of the jth branch pipeline, respectively.

2.1.3.2 Temperature Drop Model

The temperature drops exponentially along water flow in a pipe. The outlet water temperature of a pipe can be approximately determined by (2.10) (Awad et al. 2009)

$$T_{\mathrm{op}} = \begin{cases} \left(T_{\mathrm{ip}} - T_{\mathrm{a}}\right)(1-\gamma) + T_{\mathrm{a}} & \gamma \le 1 \\ T_{\mathrm{a}} & \gamma > 1 \end{cases} \tag{2.10}$$

where $\gamma = \dfrac{K_{\mathrm{p}} l}{c_{\mathrm{w}} M_{\mathrm{p}}}$, T_{op}, T_{ip}, M_{p}, K_{p} and l are the outlet water temperature, inlet water temperature, water flow rate, heat transfer coefficient and length of a pipe, respectively, and T_{a} is the ambient temperature.

2.1.4 *Heating Load*

2.1.4.1 Load Calculation

Each building is regarded as a single block from the viewpoint of a heating network. The heating load of building j is estimated based on the volumetric heat index method by (2.11)

$$Q_{\mathrm{HL}\,j} = q_{\mathrm{V}\,j} V_j \left(T_{\mathrm{n}\,j} - T_{\mathrm{a}}\right) \tag{2.11}$$

where subscript j represents building j $(j = 1, 2, \ldots, n)$, $Q_{\mathrm{HL}\,j}$, $q_{\mathrm{V}\,j}$, V_j and $T_{\mathrm{n}\,j}$ are the heating load, volumetric heat index, peripheral volume and indoor temperature of building j, respectively.

2.1.4.2 Heating Radiator Model

The model of a heating radiator presented in Jiang et al. (2014) is utilized and the heat release rate is approximately calculated as follows:

$$\begin{aligned} Q_{\mathrm{rad}\,j} &= K_{\mathrm{r}\,j} A_{\mathrm{r}\,j} \left(\frac{T_{\mathrm{w,in}\,j} + T_{\mathrm{w,out}\,j}}{2} - T_{\mathrm{n}\,j}\right) \\ &= a_j A_{\mathrm{r}\,j} \left(\frac{T_{\mathrm{w,in}\,j} + T_{\mathrm{w,out}\,j}}{2} - T_{\mathrm{n}\,j}\right)^{1+b_j} \end{aligned} \tag{2.12}$$

where $Q_{\mathrm{rad}\,j}$, $T_{\mathrm{w,in}\,j}$, $T_{\mathrm{w,out}\,j}$, $K_{\mathrm{r}\,j}$ and $A_{\mathrm{r}\,j}$ are the heat release rate, the inlet water temperature, the outlet water temperature, heat transfer coefficient, and the total heat radiating area of the radiators installed in building j, respectively, and a_j and b_j are experimentally determined parameters related to $K_{\mathrm{r}\,j}$.

The heat transferred from the network to the radiators is determined by (2.13)

$$Q_{\mathrm{net}\,j} = M_{\mathrm{w}\,j} c_{\mathrm{w}} \left(T_{\mathrm{w,in}\,j} - T_{\mathrm{w,out}\,j} \right) \tag{2.13}$$

where $Q_{\mathrm{net}\,j}$ is the heat transferred from the network to the radiators installed in building j. When the system reaches steady-state conditions, the following balance equation can be obtained:

$$Q_{\mathrm{HL}\,j} = Q_{\mathrm{rad}\,j} = Q_{\mathrm{net}\,j}. \tag{2.14}$$

2.1.5 *Cooling Load*

2.1.5.1 Load Calculation

The total cooling load of a building, including sensible and latent load, is caused by heat transferred through the building envelope (walls, roof, floor, windows, doors, *etc.*) and heat generated by occupants, equipment, and lights. The sensible load affects the dry bulb temperature, while the latent load affects the moisture content of the conditioned space. In this research, the cooling load is calculated using the CLTD/SCL/CLF method (Spitler et al. 1993) based on the designed data and hourly based weather data of a typical summer day.

The factor values of CLTD, CLF and SCL can be obtained from tables presented in AHSRAE fundamentals handbook (ASHRAE 1997). Since the ASHRAE tables provide hourly CLTD values for one typical set of conditions, *i.e.*, indoor designed temperature of 78 °F, outdoor maximum temperature of 95 °F with mean temperature of 85 °F, and daily range of 21 °F. Thus, the CLTD values should be corrected slightly in other conditions. The correction of CLTD is expressed as

$$\mathrm{CLTD}_{\mathrm{corrected}} = \mathrm{CLTD}_{\mathrm{table}} + [(1.8T_{\mathrm{mean}} + 32) - 85] - [(1.8T_{\mathrm{n}} + 32) - 78] \tag{2.15}$$

where $\mathrm{CLTD}_{\mathrm{corrected}}$ is the corrected CLTD value, $\mathrm{CLTD}_{\mathrm{table}}$ is the CLTD value obtained from ASHRAE table, T_{n} is indoor temperature, and T_{mean} is mean outdoor temperature.

2.1.5.2 AHU Model

To predict the required chilled water flow rate of each building under different operating conditions, an air handling unit (AHU) model is assumed and used to represent all terminal units in the cooling mode. It can be concluded from (Ma and Wang 2009; Ni et al. 2006) that the required water flow rate for a given condition is strongly dependent on the cooling load, air flow rate, and AHU inlet air and water temperatures. An empirical formula presented in (Ma and Wang 2009) is used to predict the performance variations of an AHU associated with the changes of major variables, which is given by

$$M_{\mathrm{w}} = d_0 \left(Q_{\mathrm{CL}}\right)^{d_1} \left(M_{\mathrm{a}}\right)^{d_2} \left(h_{\mathrm{a,in}} - h_{\mathrm{w,in}}\right)^{d_3} \tag{2.16}$$

where d_0-d_3 are coefficients, Q_{CL} is the total cooling load of a building, M_{w} and M_{a} are the chilled water flow rate and the air flow rate of an AHU, respectively, $h_{\mathrm{a,in}}$ is the specific enthalpy of air at the inlet of AHU, $h_{\mathrm{w,in}}$ is the specific enthalpy of saturated air, whose temperature is the same as the AHU inlet water temperature.

2.2 Combined Cooling Heating and Power Systems

2.2.1 *Description of CCHP Microgrid*

Based on our previous work (Jing et al. 2014), a more detailed system is proposed in Fig. 2.4. The model consists of solar-powered water heaters (SH), wind turbines (WT), gas-fired boilers (GB), electric water boilers (EB), gas turbines (GT) with heat recovery steam generator (HRSG), absorption chillers (AC), reciprocating chillers (RC), and load center (LC) with HC.

The system mainly works in two modes corresponding to one in winter as heating mode, and the other in summer as cooling mode. In the heating mode, the magnet valves V1 and V2 are open, while the magnet valves V3, V4, V5, and V6 are closed. During this season, the chillers are not used and the system supplies power and heating energy to the users at demand side. It is necessary to note that we assume that excess electricity produced in the system cannot be sold back to the grid. Therefore, surplus electricity may be wasted. In the cooling mode, the magnet valve V1 and V2 are closed, while magnet valves V3, V4, V5, and V6 are open. The hot water is converted to chilled water to meet user's cooling energy demand.

2.2.2 *Wind Turbine Generator*

A grid-connected WT with a permanent magnet synchronous generator (PMSG) is applied to generate electricity. The energy obtained from WT can be calculated as follows:

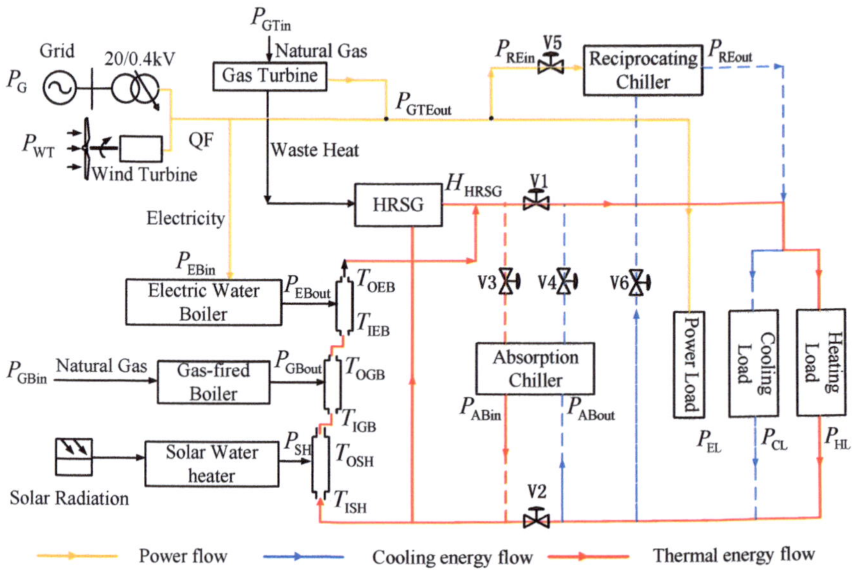

Fig. 2.4 Schematic diagram of CCHP microgrid

$$P_{WT}(t) = \begin{cases} 0 & v < v_{ci} \\ 0.5C_p\rho_0\pi r^2 v^3 & v_{ci} < v < v_{cr} \\ P_r & v_{cr} < v < v_{co} \\ 0 & v > v_{co} \end{cases} \tag{2.17}$$

where P_{WT} is the output power of WT, P_r is the rated output power, C_p is the power coefficient, ρ_0 is the density of outdoor air, v_{ci}, v_{co}, and v_r are the cut-in, cut-out, and rated speed of WT, respectively.

2.2.3 *Gas Turbine and Heat Recovery Steam Generator*

The mathematical model of GT can be described by the following equations:

$$\begin{aligned} P_{GTin} &= B_g V_{gGT} \\ H_{HRSG} &= \eta_{HRSG} P_{GTin} \\ P_{GTout} &= \eta_{GT} P_{GTin} = \dot{M}_2 c_W (T_{OGT} - T_{IGT}) \end{aligned} \tag{2.18}$$

where V_{GT} is the natural gas consumption rate per second of the GT.

2.2.4 *Load Center*

The LC is equipped with both household air conditioners, which can transform electricity to cooling energy or heating energy. The energy equality constrains of LC can be described as

$$\begin{aligned} D_{\mathrm{p}} &= P_{\mathrm{pp}} + P_{\mathrm{pc}} + P_{\mathrm{ph}} \\ D_{\mathrm{c}} &= P_{\mathrm{cl}} + P_{\mathrm{pc}} * \eta_{\mathrm{EC}} \\ D_{\mathrm{h}} &= P_{\mathrm{hl}} + P_{\mathrm{ph}} * \eta_{\mathrm{EH}} \end{aligned} \quad (2.19)$$

where P_{pc} and P_{ph} are electricity consumed by electric conditioner to generate cooling and heating energy, respectively. P_{pp} is electricity consumed by other electrical appliances.

2.3 Energy Hubs

2.3.1 *System Description*

The framework of typical EH presented in this section is illustrated in Fig. 2.5. The EH includes energy supply terminals, energy conversion equipment terminals, and end user terminals. As we can see in Fig. 2.1, utility grid, photovoltaic generation, and natural gas form the energy supply terminals. As for energy conversion equipment terminals, it consists of several component like gas turbine (GT), PV panel, energy storage battery (ESB), heat recovery steam generator (HRSG) with afterburning (AB), absorption chiller (AC), heat exchanger (HE), and electric chiller (EC). Besides, the end user terminals are composed of the electric load demand, hot water demand, and cooling demand.

In EH, the GT, which can generate electricity and recoverable heat simultaneously, is applied as power generator unit in the system. On the one hand, the electricity supply needs to not only satisfy part electricity demand but also drive the EC, while the electricity power provided by the GT is insufficient, the shortage part will be

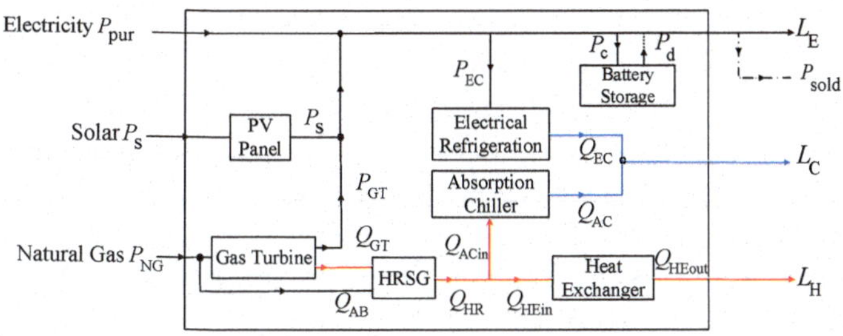

Fig. 2.5 Schematic diagram of an EH

supplied by PV panel and public grid. Conversely, when the electricity power GT provided exceeds the requirement, the excess electricity power will be stored in the battery or sold back to the public grid. On the other hand, recovered heat is divided into two parts.

One part is transformed by the HE to meet the hot water requirement of end users. The other part is applied to drive the AC to work normally to generate the required cooling load. Once the hot water obtained by heat exchanger can not completely meet the hot water requirement, the rest requirement will be provided by the HRSG. In terms of the cooling requirement, it can be gained from two ways: EC and AC, respectively.

2.3.2 *Gas Turbine*

GT, which provides electricity and recoverable heat simultaneously, is regarded as a power generation unit.

when considering the off-design characteristic of GT, the output of electricity, the amount of thermal output, and the amount of gas fuel consumption have a coupling relationship. And the output of electricity, the gas fuel consumption are completely determined by the thermal output. Hence, the characteristic performance of GT based on off-design model is expressed below (Li et al. 2003):

$$P_{\mathrm{GT}}(t) = K_1 Q_{\mathrm{GT}}^2(t) + K_2 Q_{\mathrm{GT}}(t) + K_3 \tag{2.20}$$

$$P_{\mathrm{NG}}(t) = K_4 Q_{\mathrm{GT}}^2(t) + K_5 Q_{\mathrm{GT}}(t) + K_6 \tag{2.21}$$

$$V_{\mathrm{GT}}(t) = \frac{P_{\mathrm{NG}}(t)}{\mathrm{LHV}} \tag{2.22}$$

where NG refers to natural gas, LHV indicates lower heating value.

2.3.3 *Energy Storage Battery*

For the purpose of shaving the peak demand and compensating the fluctuation of renewable energy, ESB is employed in the EH.

What's more, in this chapter, we choose the Lithium-ion (Li-ion) battery because of its optimal energy-to-weight ratio and the slow loss of charge (Gao et al. 2002). The state of charge (SOC) is its unique state variable.

Since charging and discharging cannot be performed simultaneously, $P_c(t)$ and $P_d(t)$ have the mutually exclusive relationship which can be expressed through two binary variables $\gamma_{ESB.t}^{ch}$ and $\gamma_{ESB.t}^{dis}$. In this way, the equations describing charge and discharge can be formulated below (Chen et al. 2012):

$$SOC(t+1) = (1-f)SOC(t) + \frac{\eta_c P_c(t)}{E}\gamma_{ESB.t}^{ch} - \frac{P_d(t)}{\eta_d E}\gamma_{ESB.t}^{dis}$$
$$\gamma_{ESB.t}^{ch} + \gamma_{ESB.t}^{dis} = 1 \tag{2.23}$$
$$\forall t \in T$$

where T is total number of hours per day (24 h).

2.3.4 *Heat Recovery Steam Generator with Afterburning*

The HRSG, which has the ability to significantly improve the energy efficiency by recovering the high-grade waste heat, is more and more widely applied in cogeneration cycle systems (Liu et al. 2004). Generally speaking, HRSG turns the room temperature water into hot water we need by recovering the waste heat generated from the GT. During shutdown times of the GT, the hot water demand is met by the thermal of complemental burning of natural gas. In actual operation, owing to the difference in users' need for steam temperature, HRSG is often operated under different parameters. Therefore, The efficiency and rated value of the HRSG is usually different, which will affect the operation performance of the entire EH.

In order to simplify the complexity of the model, in this article, we consider the thermal characteristics of the HRSG to be related to its rated thermal output Q_{HRN} and its rated thermal efficiency η_{HRN}. The relationship of the Q_{HRN} and η_{HRN} can be described as (Li et al. 2003)

$$\frac{\eta_{\mathrm{HR}}(t)}{\eta_{\mathrm{HRN}}} = K_7\left(\frac{Q_{\mathrm{HR}}(t)}{Q_{\mathrm{HRN}}}\right)^2 + K_8\left(\frac{Q_{\mathrm{HR}}(t)}{Q_{\mathrm{HRN}}}\right) + K_9 \tag{2.24}$$

$$Q_{\mathrm{HRin}}(t) = Q_{\mathrm{AB}}(t) + Q_{\mathrm{GT}}(t) \tag{2.25}$$

$$Q_{\mathrm{AB}}(t) = V_{\mathrm{AB}}(t)\mathrm{LHV} \tag{2.26}$$

2.3.5 *Absorption Chiller*

The AC has been studied by the majority of researchers due to its reliability and full utilization in the low- grade heat of waste heat. Until now, there are three types of absorption chillers for providing users with cooling requirement: single-, double-, and triple-effect chillers. Furthermore, the temperature of heat resource is the key factor to be considered when choosing the chiller. The thermal characteristics performance of the AC can be presented as follows when the off-design characteristics of the AC is considered in the model (Li et al. 2006):

$$\frac{\mathrm{COP}_{\mathrm{AC}}(t)}{\mathrm{COP}_{\mathrm{ACN}}}=\frac{\frac{Q_{\mathrm{AC}}(t)}{Q_{\mathrm{ACN}}}}{K_{10}\left(\frac{Q_{\mathrm{AC}}(t)}{Q_{\mathrm{ACN}}}\right)^{3}+K_{11}\left(\frac{Q_{\mathrm{AC}}(t)}{Q_{\mathrm{ACN}}}\right)^{2}+K_{12}\left(\frac{Q_{\mathrm{AC}}(t)}{Q_{\mathrm{ACN}}}\right)+K_{13}} \tag{2.27}$$

$$Q_{\mathrm{AC}}(t)=\mathrm{COP}_{\mathrm{AC}}(\mathrm{t})\mathrm{Q}_{\mathrm{ACin}}(\mathrm{t}) \tag{2.28}$$

where ACN is defined as input of absorption chiller.

2.3.6 *Electric Chiller*

Unlike the AC which can only be driven by the low-quality waste heat, the EC can generate cooling by means of consuming electricity. Therefore, it is obvious that the COP of EC is higher than AC on account of the consumption of high-quality electricity. The thermal performance of EC is formulated as follows when considering the off-design characteristics of EC (Li et al. 2006):

$$\frac{\mathrm{COP}_{\mathrm{EC}}(t)}{\mathrm{COP}_{\mathrm{ECN}}}=\frac{\frac{Q_{\mathrm{EC}}(t)}{Q_{\mathrm{ECN}}}}{K_{14}\left(\frac{Q_{\mathrm{EC}}(t)}{Q_{\mathrm{ECN}}}\right)^{2}+K_{15}\left(\frac{Q_{\mathrm{EC}}(t)}{Q_{\mathrm{ECN}}}\right)+K_{16}} \tag{2.29}$$

$$Q_{\mathrm{EC}}(t)=\mathrm{COP}_{\mathrm{EC}}(t)P_{\mathrm{EC}}(t) \tag{2.30}$$

2.3.7 *Heat Exchanger*

Since the thermal efficiency of HE is basically the same under different thermal load conditions, therefore in this section we considered a fixed thermal efficiency which is represented as

$$\eta_{\mathrm{HE}}(t)=\eta_{\mathrm{HEN}}=\frac{Q_{\mathrm{HEout}}(t)}{Q_{\mathrm{HEin}}(t)} \tag{2.31}$$

where $\mathrm{HE}_{\mathrm{in}}$ and $\mathrm{HE}_{\mathrm{out}}$ represent the input and output of heat exchanger, respectively.

2.3.8 *Photovoltaic Generation*

Solar photovoltaic systems convert solar irradiation into electricity, which is needed for photoelectric effects. The basic building block of solar photovoltaic power is either solar cells or photovoltaic cells (Omer 2008). In this section, photovoltaic cell is operated in accordance with the predictive output $P_{\mathrm{S}}(t)$.

2.4 Individual-Based Model for the LSIES

2.4.1 Definition of IBM

A compatible unified model for complex system, called individual-based model (IBM) is proposed. IBM is composed of individual models Σ and system models M.

An individual comprises of a quinary tuple of input, knowledge, function, state, and output sets. A standard description of the basic IBM is given by

$$\Sigma = (\boldsymbol{I}, \boldsymbol{K}, \boldsymbol{X}, \boldsymbol{F}, \boldsymbol{O}) \tag{2.32}$$

where Σ is a mathematical description of an individual which can be described by a quinary tuple. $\boldsymbol{I}$ is the input set, $\boldsymbol{O}$ is the output set, $\boldsymbol{K}$ is the knowledge set, $\boldsymbol{X}$ is the state set, and $\boldsymbol{F}$ is the function set. Layout of internal mechanism of an IBM is provided in Fig. 2.6

- The input of an individual at time t, denoted by i_t, is given by (2.33)

$$i_t = \{(i, t)|i \in \boldsymbol{I}, t \in T\} \tag{2.33}$$

where T is the time sequence of input and output information from the environment. $\boldsymbol{I}$ is input set which changes during a finite time interval. The information of the environment can be transmitted in through this set.

$$\boldsymbol{I} = \{i_1, i_2, \ldots, i_t\} \tag{2.34}$$

- The inherent knowledge of an individual at time t, denoted by k_t, is given by (2.35):

$$k_t = \{(k, t)|k \in \boldsymbol{K}, t \in T\} \tag{2.35}$$

where $\boldsymbol{K}$ is the set of individual inherent attributes.

$$\boldsymbol{K} = \{k_1, k_2, \ldots, k_t\} \tag{2.36}$$

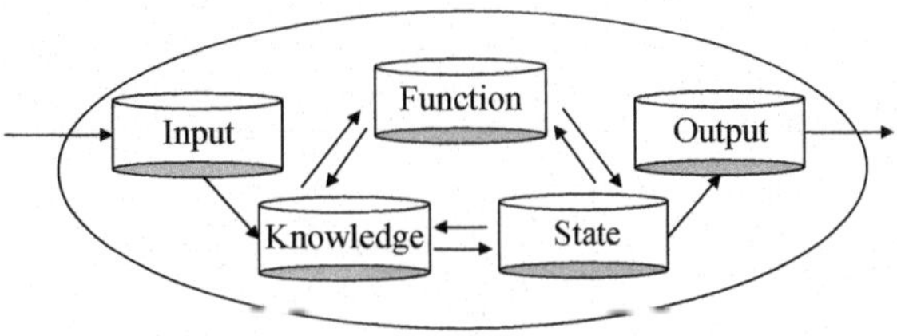

Fig. 2.6 Layout of internal mechanism of an IBM

The inherent knowledge usually composes of parameters, attributes, and structures of an individual. The knowledge is continuously modified along the time as required. For example, k_1 can be different from k_t under the circumstance that fault causing structural changes of the system.

- $\boldsymbol{X}_t$ is the state of an individual at time t, which is an element of state set X.

$$x_t = \{(x, t)|x \in \boldsymbol{X}, t \in [0, e_t]\} \tag{2.37}$$

where e_t is the maximum time that capture the dynamic process of the system (Fernando 2002).

$$\boldsymbol{X} = \{X_1, X_2, \ldots, X_t\} \tag{2.38}$$

The state set records individual status which characterizes the dynamic process of the individuals.

- $\boldsymbol{F}(\cdot, \cdot, \cdot)$ is the transition function, which describes certain behaviors and mechanisms of an individual. It is a mapping from $X_t \times I_t \rightarrow X_{t+1}$. In this part, self-evolution occurs spontaneously at time $t + 1$ according to the state, knowledge, and input at time t. With this function, a system changes its state autonomously without receiving any external stimulus (Liao and Wu 2013).

$$X_{t+1} = F[X_t, k_t, i_t] \tag{2.39}$$

Note that the function set is not only composed of differential equations on physical regulation and constraints, but also unquantified individual rules.

- o_t is the individual output at time t, is given by (2.40)

$$o_t = \{(o, t)|i \in \boldsymbol{O}, t \in T\} \tag{2.40}$$

The specific information invoked by the environment will be updated in $\boldsymbol{O}$.

$$\boldsymbol{O} = \{o_1, o_2, \ldots, o_t\} \tag{2.41}$$

Throughout this process, all information of the individual is hidden from the environment except information in $\boldsymbol{O}$. In this way, information privacy of the individuals can be protected. Also, autonomy in $\boldsymbol{F}$ means independent operation without intervention from the environment.

The fully specified IBM consists of layered objects. One individual as the most basic set is a part of a small system as part of a large system. Figure 2.7 expresses a three-stage system and its mathematical model is depicted in (2.42).

$$\begin{aligned} &\Sigma_i = (I_i, K_i, X_i, F_i, O_i) && i = 1, 2, 3, \ldots, N_{L1} \\ &\Sigma_{ij} = (I_{ij}, K_{ij}, X_{ij}, F_{ij}, O_{ij}) && j = 1, 2, 3, \ldots, N_{L2} \\ &\Sigma_{ijk} = (I_{ijk}, K_{ijk}, X_{ijk}, F_{ijk}, O_{ijk}) && k = 1, 2, 3, \ldots, N_{L3} \end{aligned} \tag{2.42}$$

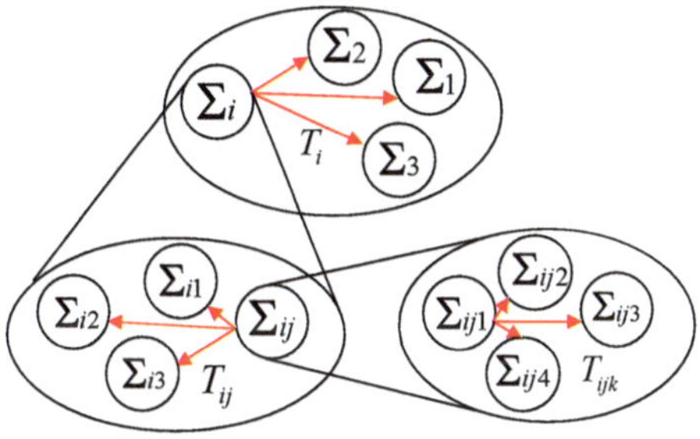

Fig. 2.7 IBM of a hierarchical system

where Σ_i, Σ_{ij}, and Σ_{ijk} represents for the large systems, small systems, and basic units, respectively. $N_{\mathrm{L1}} \sim N_{\mathrm{L3}}$ represent for the numbers of IBMs on each level, respectively.

A system model M is further expressed as

$$M = (V, T) \tag{2.43}$$

where $V = \{\Sigma | \Sigma \in V\}$, which is the sets of subindividuals. T is the incidence matrix applied to describe the connections between the subindividuals. The constituent element T_{ij} is evaluated by the following rules:
$+1$, if the output of Σ_i flows to the input of Σ_j;
-1, if the output of Σ_j flows to the input of Σ_i;
0, if there is no connection between the two individuals.

From this figure, we can see that in the hierarchical philosophy of IBM, a large system contains several small systems and each small system contains basic IBMs, and a complex system is fully described by modeling heterogeneous individuals by their own rules. The individuals with various characteristics inside communicate with others through input and output sets in a unified form. Furthermore, given the fact that each individual is independent with respect to others, parallel activities can be processed in different parts of the system.

2.4.2 *IBM for the LSIES*

2.4.2.1 Descriptions of LSIES

LSIES which is composed of a WSCC 3-generator-9-bus Hughes (1977) EPS and 6-node HS is modeled in this Section. There are two coupling units in the LSIES: CHP 1 integrates generator 1 in EPS and thermal source 1 in HS. As an extraction steam turbine CHP unit operating in the following electric load (FEL) mode, greater electricity generation leads to more thermal energy. CHP 2 integrates generator 3 in EPS and thermal source 2 in HS. As a gas turbine CHP unit operating in the following

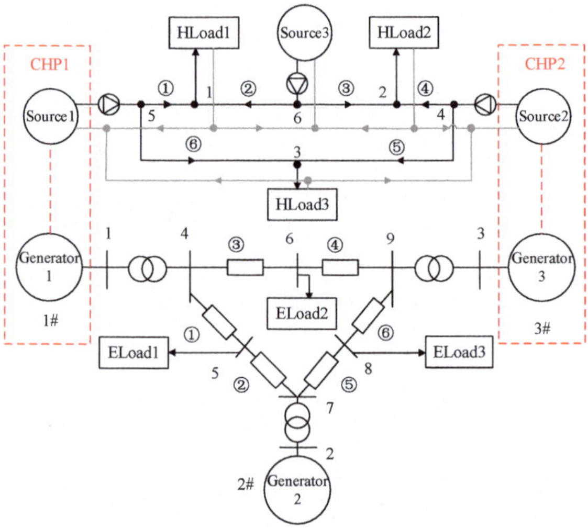

Fig. 2.8 The schematic diagram of 6-node HS

thermal load (FTL) mode, greater thermal energy generation leads to lower electricity generation (Pan et al. 2016).

Figure 2.8 is the outline of IES and the corresponding IBM model is shown in Fig. 2.9. There are four types of basic models in IES: the load center individuals, the generator individuals, the source individuals, and the network individuals. Different components inherit different basic models and makeup EPS and HS, as parts of

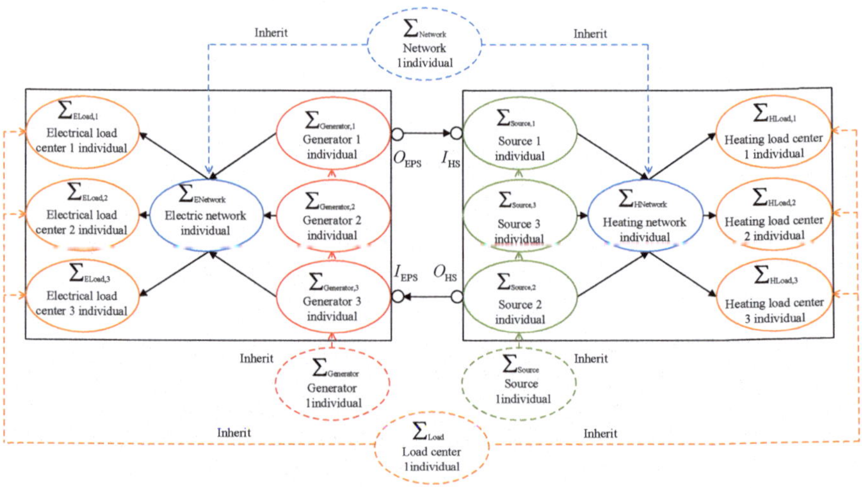

Fig. 2.9 The schematic diagram of LSIES

LSIES. The subindividuals interact with others in a unified manner through input and output sets. In this section, generalized energy flow is adopted to express the correlations.

Dynamic analysis of LSIES is divided into several steps, note that the IBM for HS is dynamically changing in different steps. Step 0 is when IES is stable and both of the individuals are in a normal stage. And then a disturbance occurs and followed by step 1 within several seconds, the generator transient stage. In this step, most variables in HS such as the mass flow rate and water temperature in the heating network remain unchanged. Nevertheless, the transient process in EPS occurs and the power outputs of generators are not the same with step 0. Afterward, the hydraulic process of HS happens in a few seconds, followed by step 2, the hydraulic stage. The mass flow rate of the water in the pipeline in this stage is no longer the same with step 1, although the water temperature still remains unchanged. With the hot water running, step 3 comes within minutes. In this stage, the power flow in EPS, the mass flow rate and water temperature in HS all change compared with the last step. In the following steps, the status of one individual change because of the disturbance transformed by the coupling components from the other individual until the whole system reaches a steady state.

2.4.2.2 IBM for EPS

EPS is composed of several individuals such as the generators, the network, and the load centers. All of the individuals can be described separately by IBM independently. Due to the space limitations, only the IBM for the integrated EPS is illustrated in (2.44).

$$\begin{aligned} \Sigma\text{EPS} &= (I_{\text{EPS}}, K_{\text{EPS}}, X_{\text{EPS}}, F_{\text{EPS}}, O_{\text{EPS}}) \\ I_{\text{EPS}} &= \{P_{\text{G},3}\} \\ K_{\text{EPS}} &= \{K_{\text{Eload}}, K_{\text{ENetwork}}, K_{\text{Generator}}\} \\ X_{\text{EPS}} &= \{P, Q, V\} \\ F_{\text{EPS}} &= \{(2.45) \sim (2.47)\} \\ O_{\text{EPS}} &= \{P_{\text{G},1}\} \end{aligned} \tag{2.44}$$

where K_{Eload}, K_{ENetwork}, and $K_{\text{Generator}}$ are the knowledge sets of electrical load centers, electric network, and the generators, respectively. Parameter settings of the corresponding individual can be found in these sets.

- Electrical load center: the electrical load center models are indicated as the fixed active and reactive power demand at bus i. $K_{\text{Eload}} = \{P_{\text{Load}}, Q_{\text{Load}}\}$.
- Generator: the fourth-order model is employed for the synchronous generator, in which a IEEE DC-I exciter and a PSS is adopted for the control of synchronous generators, the details can be found in Liu et al. (2014). $K_{\text{Generator}} = \{X_{\text{d}}, X'_{\text{d}}, X_{\text{q}}, X'_{\text{q}}, T'_{\text{d0}}, T'_{\text{q0}}, E'_{\text{f}}, T_{\text{J}}, T_{\text{m}}\}$. The generator individuals are expressed as

$$
\begin{aligned}
&T'_{\mathrm{d0}}\frac{\mathrm{d}E'_{\mathrm{q}}}{\mathrm{d}t} = E'_{\mathrm{f}} - E'_{\mathrm{q}} - (X_{\mathrm{d}} - X'_{\mathrm{d}})i_{\mathrm{d}} \\
&T'_{\mathrm{q0}}\frac{\mathrm{d}E'_{\mathrm{d}}}{\mathrm{d}t} = -E_{\mathrm{d}} = -E'_{\mathrm{d}} + (X_{\mathrm{q}} - X'_{\mathrm{q}})i_{\mathrm{q}} \\
&T_{\mathrm{J}}\frac{\mathrm{d}\omega}{\mathrm{d}t} = T_{\mathrm{m}} - T_{\mathrm{e}} = T_{\mathrm{m}} - [E'_{\mathrm{q}}i_{\mathrm{q}} + E'_{\mathrm{d}}i'_{\mathrm{d}} - (X'_{\mathrm{d}} - X'_{\mathrm{q}})i_{\mathrm{d}}i_{\mathrm{q}}] \\
&\frac{\mathrm{d}\delta}{\mathrm{d}t} = \omega - 1
\end{aligned}
\tag{2.45}
$$

- Electric network: $K_{\mathrm{ENetwork}} = \{G_{ij}, B_{ij}\}$.

$$
\begin{aligned}
P_{\mathrm{G}_i} &= P_{\mathrm{D}_i} + V_i \sum_{j\in N_i} V_j(G_{ij}\cos\theta_{ij} + B_{ij}\sin\theta_{ij}) \\
Q_{\mathrm{G}_i} &= Q_{\mathrm{D}_i} + V_i \sum_{j\in N_i} V_j(G_{ij}\sin\theta_{ij} - B_{ij}\cos\theta_{ij})
\end{aligned}
\tag{2.46}
$$

- Disturbances: To specify the disturbances and faults occurred in EPS, unquantified rules are included within the IBM model, which are expressed as

$$
\begin{aligned}
&\textbf{if } \alpha = 1 \textbf{ then} \\
&\quad \text{when } t \geq T_{\mathrm{Dis}},\ P_{\mathrm{Load}} = \hat{P}_{\mathrm{Load}},\ Q_{\mathrm{Load}} = \hat{Q}_{\mathrm{Load}}\ ; \\
&\textbf{else if } \alpha = 2 \textbf{ then} \\
&\quad \text{when } T_{\mathrm{Dis}} \leq t < T_{\mathrm{Nor}},\ G = \hat{G},\ B = \hat{B}; \\
&\quad \text{when } t \geq T_{\mathrm{Nor}},\ \hat{G} = G,\ \hat{B} = B; \\
&\textbf{end if}
\end{aligned}
\tag{2.47}
$$

  where α is the disturbance type, and "1" represents for load variation and "2" represents for fault. T_{Dis} and T_{Nor} are time when the disturbance occurs and cleared, respectively. $\hat{P}_{\mathrm{Load}}$, $\hat{Q}_{\mathrm{Load}}$ are active and reactive power demand after disturbances, respectively. $\hat{G}$ and $\hat{B}$ are network parameters after disturbances.

The EPS model in this section is implemented by a Matlab/Simulink based power system simulation toolbox (PSST) proposed by Yao, more details can be found in Yao et al. (2012).

2.4.2.3 IBM for HS

For HS, the quasi-steady model is applied to describe its hydraulic and thermal characteristics, and the details of the model have been provided in Wang et al. (2018). The IBM of HS is expressed in (2.48).

$$
\begin{aligned}
\Sigma_{\mathrm{HS}} &= (I_{\mathrm{HS}}, K_{\mathrm{HS}}, X_{\mathrm{HS}}, F_{\mathrm{HS}}, O_{\mathrm{HS}}) \\
I_{\mathrm{HS}} &= \{P_{\mathrm{G},1}\} \\
K_{\mathrm{HS}} &= \{K_{\mathrm{Hload}}, K_{\mathrm{HNetwork}}, K_{\mathrm{Source}}\} \\
X_{\mathrm{HS}} &= \{T, m, \Phi\} \\
F_{\mathrm{HS}} &= \{(2.49) \sim (2.53)\} \\
O_{\mathrm{HS}} &= \{P_{\mathrm{G},3}\}
\end{aligned}
\tag{2.48}
$$

where K_{Hload}, K_{HNetwork}, and K_{Source} are the knowledge sets of thermal load centers, thermal network, and the sources, respectively. Parameter settings of the corresponding individual can be found in these sets.

- Thermal load center: the thermal load center models are indicated as the fixed thermal demand at bus i. $K_{\mathrm{Hload}} = \{\Phi_{\mathrm{Load}}\}$.
- Thermal network: $K_{\mathrm{HNetwork}} = \{\boldsymbol{A}, \boldsymbol{B}, L, T_{\mathrm{a}}, T_{\mathrm{source}}, k_{\mathrm{p}}, \lambda, c_{\mathrm{w}}\}$.
  The hydraulic models of heating network:

$$
\begin{aligned}
\boldsymbol{A}m &= m_{\mathrm{q}} \\
\boldsymbol{B}h_{\mathrm{f}} &= 0 \\
h_{\mathrm{f}} &= k_{\mathrm{p}}m|m|
\end{aligned}
\tag{2.49}
$$

  where $\boldsymbol{A}$,$\boldsymbol{B}$ are the network incidence matrix and loop incidence matrix of HS individual, respectively (Liu et al. 2014).
  The thermal models of heating network:

$$
\begin{aligned}
\Phi &= c_{\mathrm{w}}m(T_{\mathrm{r}} - T_{\mathrm{s}}) \\
T_{\mathrm{se}} &= (T_{\mathrm{s}} - T_{\mathrm{a}})(1 - \frac{\lambda L}{c_{\mathrm{w}}m}) + T_{\mathrm{a}} \\
(\Sigma m_{\mathrm{out}})T_{\mathrm{out}} &= \Sigma(m_{\mathrm{in}}T_{\mathrm{in}})
\end{aligned}
\tag{2.50}
$$

- Source: $K_{\mathrm{HNetwork}} = \{U, k_{\mathrm{HP}}\}$.

$$
\Phi = U - \tfrac{1}{k_{\mathrm{HP}}}P_{\mathrm{G}}
\tag{2.51}
$$

  where U is coefficient of power heat output of CHP, and $\mathrm{k_{HP}}$ is thermal to power ratio of CHP.
- Disturbances: Unquantified rules which specify the disturbances and faults occurred in HS are expressed as

$$
\begin{aligned}
&\textbf{if } \alpha = 1 \textbf{ then} \\
&\quad \text{when } t \geq T_{\mathrm{Dis}},\ \Phi_{\mathrm{Load}} = \hat{\Phi}_{\mathrm{Load}}; \\
&\textbf{else if } \alpha = 2 \textbf{ then} \\
&\quad \text{when } T_{\mathrm{Dis}} \leq t < T_{\mathrm{Nor}},\ \boldsymbol{A} = \hat{\boldsymbol{A}},\ \boldsymbol{B} = \hat{\boldsymbol{B}}; \\
&\quad \text{when } t \geq T_{\mathrm{Nor}},\ \hat{\boldsymbol{A}} = \boldsymbol{A},\ \hat{\boldsymbol{B}} = \boldsymbol{B}; \\
&\textbf{end if}
\end{aligned}
\tag{2.52}
$$

where $\hat{\Phi}_{\text{Load}}$ is thermal energy demand after disturbances. $\hat{\boldsymbol{A}}$ and $\hat{\boldsymbol{B}}$ are network parameters after disturbances.

- Stages: Unquantified rules which specify the stages of HS are expressed as

$$
\begin{aligned}
&\textbf{if } 0 \le t < T_{\text{Hydr}} \textbf{ then}\\
&\quad m_t = m_{t-1},\ T_t = T_{t-1};\\
&\textbf{else if } T_{\text{Hydr}} \le t < T_{\text{Ther}} \textbf{ then}\\
&\quad T_t = T_{t-1};\\
&\textbf{end if}
\end{aligned}
\tag{2.53}
$$

where T_{Hydr} and T_{Ther} are time of hydraulic and thermal stages of HS.

2.4.2.4 IBM for LSIES

The system model for LSIES is illustrated as follows:

$$\Sigma_{\text{IES}} = (V_{\text{IES}}, T_{\text{IES}}) \tag{2.54}$$

where $V_{\text{IES}} = \{\Sigma_{\text{EPS}}, \Sigma_{\text{HS}}\}$, $T_{\text{IES}} = [0\ 1;\ 1\ 0]$.

2.5 Summary

In this chapter, we establish the models of district heating and cooling systems, combined cooling heating and power systems and energy hubs for analysis of LSIES. Furthermore, an individual-based model (IBM), composed of individual models and heterogeneous system models, is proposed for modeling and analysis of LSIES. A quintuple of input, knowledge, state, function, and output sets is expressed to illustrate a basic individual. Particular evolutionary mechanisms in the function set of the two subindividuals, EPS and HS, ensure the models be described accurately with its own particular characteristics. Furthermore, the two individuals interact with others in a unified manner with generalized energy flow. Dynamic analysis of IES under four types of disturbances and faults occurring in different individuals are investigated. The IBMs are capable of simulating quantified system states and unquantified rules simultaneously. The unquantified rules are modeled using the function set to depict the relationships between different variables. More importantly, the IBMs are in an open architecture which can be used to aggregate various individuals modeled by different developers.

References

ASHRAE Handbook of Fundamentals (1997) chap 27. ASHRAE

Awad B, Chaudry M, Wu JZ, Jenkins N (2009) Integrated optimal power flow for electric power and heat in a microgrid. In: Proceedings of 20th International Conference and Exhibition on Electricity Distribution, pp 1–4. Prague, Czech republic

Chen S, Gooi HB, Wang MQ (2012) Sizing of energy storage for microgrids. In: Power and Energy Society General Meeting, pp 1–1

Dolan PS, Nehrir MH, Gerez V (1996) Development of a Monte Carlo based aggregate model for residential electric water heater loads. Electr Power Syst Res 36(1):29–35

Fernando JB (2002) Modeling and simulation of dynamic structure heterogeneous flow systems. Simul J Soc Comput Simul (1), 18–27

Gao L, Liu S, Dougal RA (2002) Dynamic lithium-ion battery model for system simulation. IEEE Trans Compon Packag Technol 25(3):495–505

Gordon JM, Ng KC (1994) Thermodynamic modeling of reciprocating chillers. J Appl Phys 75(6):2769–2774

Gordon JM, Ng KC (1995) A general thermodynamic model for absorption chillers: theory and experiment. Heat Recover Syst CHP 15(1):73–83

Hughes FM (1977) Power system control and stability. Iowa State University Press

Jiang XS, Jing ZX, Li YZ, Wu QH, Tang WH (2014) Modeling and operation optimization of an integrated energy based direct district water-heating system. Energy 64:375–388

Jing ZX, Jiang XS, Wu QH, Tang WH, Hua B (2014) Modelling and optimal operation of a small-scale integrated energy based district heating and cooling system. Energy 73:399–415

Li H, Nalim R, Haldi PA (2006) Thermal-economic optimization of a distributed multi-generation energy systema case study of beijing. Appl Therm Eng 26(7):709–719

Liao HL, Wu QH (2013) Multi-objective optimization by learning automata. Kluwer Academic Publishers

Li H, Haldi PA, Favre PP (2003) Evaluation of a distributed energy system combined with heating, cooling and power generation through multi-criteria optimization. In: ASME 2003 International Mechanical Engineering Congress and Exposition, pp 277–284

Liu XZ, Nick J, Wu JZ, Audrius B (2014) Combined analysis of electricity and heat networks. Energy Procedia 61, 155–159. In: International Conference on Applied Energy, ICAE2014

Liu BT, Chien KH, Wang CC (2004) Effect of working fluids on organic rankine cycle for waste heat recovery. Energy 29(8):1207–1217

Liu Y, Wu QH, Zhou XX, Jiang L (2014) Perturbation observer based multiloop control for the dfig-wt in multimachine power system. IEEE Trans Power Syst 29(6):2905–2915

Ma Z, Wang SW (2009) An optimal control strategy for complex building central chilled water systems for practical and real-time applications. Build Environ 44(6):1188–1198

Ni MQ, Liu GY, Shen W (2006) Calculating methods of cooling capacity of fan coil units under different conditions. Heat Vent Air Cond 36(8):67–69

Omer AM (2008) Drinking water from solar stills: a renewable technology for sudan. Energy Environ 19(5):657–678

Pan ZG, Guo QL, Sun HB (2016) Interactions of district electricity and heating systems considering time-scale characteristics based on quasi-steady multi-energy flow. Appl Energy 167:230–243

Paull L, Mackay D, Li H, Chang LC (2009) A water heater model for increased power system efficiency. In: Proceedings of Canadian Conference on Electrical and Computer Engineering, pp 731–734. St. Johns, NL, Canada

Spitler JD, McQuiston FC, Lindsey KL (1993) CLTD/SCL/CLF cooling load calculation method. In: Proceedings of the 1993 Winter Meeting of ASHRAE Transactions, Part 1, pp 183–192. Chicago, IL, USA

Spreitzer K, Rückbrodt D, Straky H (2002) Observer-based estimation of the water-mass-flow through a central heating boiler. In: Proceedings of American Control Conference, vol 6, pp 5054–5059. Anchorage, AK, United states

Wang LX, Jing ZX, Zheng JH, Wu QH, Wei F (2018) Decentralized optimization of coordinated electrical and thermal generations in hierarchical integrated energy systems considering competitive individuals. Energy 158:607–622

Yao W, Wen JY, Cheng SJ, Jiang L (2012) Development of a matlab/simulink based power system simulation toolbox. Power Syst Technol 36(6):95–101

Zhai XQ, Wang RZ, Dai YJ, Wu JY, Xu YX, Ma Q (2007) Solar integrated energy system for a green building. Energy Build 39(8):985–993

Chapter 3
Multi-objective Optimization Algorithms

Abstract In the LSIES, multiple benefits of different operating interests are taken into consideration. Hence, the planning and operation of LSIES are formulated as multi-objective optimization problems, which should be tackled using the multi-objective optimization algorithms. This chapter presents three multi-objective optimization algorithms, i.e., the multi-objective group search optimizer with adaptive covariance and Lévy flights (MGSO-ACL), multi-objective group search optimizer with adaptive covariance and chaotic search (MGSOACC), and multi-objective evolutionary predator and prey strategy (EPPS). Simulation studies conducted on benchmark functions are also carried out to investigate the performance of these algorithms. In later chapters, these algorithms are employed to deal with the planning and operating problems of LSIES.

Keywords Multi-objective optimization algorithms · Non-dominated sorted genetic algorithm · Multi-objective group search optimizer · Multi-objective evolutionary predator and prey strategy

3.1 Formulation of the Multi-objective Optimization Problems

3.1.1 Introduction

A many-objective optimization problems (MaOPs) is a special branch of multi-objective optimization problems (MOPs) with more than three objectives. With the multiple conflicting demands faced by the industry today for superior quality, low cost, higher safety, and so on, competitive edge could only be established by designing the products and processes that account for as many performance objectives as possible. It implies that many objectives need to be simultaneously dealt with, in an optimization problem. However, it is hard to obtain the entire solution set of a many-objective optimization problems (MaOPs) by multi-objective optimization algorithms (such as NSGA-, GSOMP) because of the difficulties brought by the curse of dimensionality (Wang and Yao 2016).

Q.-H. Wu et al., *Large-Scale Integrated Energy Systems*, Energy Systems in Electrical Engineering, https://doi.org/10.1007/978-981-13-6943-8_3

The major impediments in handling a large number of objectives are related to stagnation of search process (Deb and Saxena 2005), inefficiency of selection operators, high computational cost, and difficulty in visualization of the objective space (Saxena et al. 2013).

(1) Inefficiency of selection operators: with the increase in M, a multi-objective problem may well have a high-dimensional Pareto-set with complicated shapes, this makes most existing MOEAs and MGSO ineffective, where primary selection is based on Pareto-dominance.
(2) High computational cost: if a continuous multi-objective optimization problem (with M objectives) meets the regularity property, the dimension of its POF can be $M - 1$. Therefore, the number of points needed to approximate the whole POF increases exponentially with M. The same phenomenon can be observed in discrete problems.
(3) Difficulty in visualization of a POF for problems with $M \geq 4$: finding a higher dimensional Pareto-optimal surface is an important matter, but visualizing it for proper decision-making is equally as important.

An increasing number of related research has been put forward to overcome the difficulty of MaOPs, which can be roughly divided into three classes:

(1) Scalarization technique: all objectives are converted into a single-composite objective using the weighed sum approach. With different vectors repeating the scalarization process, there will be a number of Pareto-optimal solutions, then decision-making can be implemented to complete the optimization. However, each Pareto-optimal solution is independent of each other, thereby losing the parallel searchability often desired in solving complex optimization problems.
(2) Involvement of decision makers (DM): for a large number of objectives, it can involve a decision maker from the outset of the optimization process instead of finding the optimal solutions corresponding to a specific weight vector or varepsilon vector, although this overcomes the dimensionality problem described earlier by not finding points on the complete high-dimensional Pareto-optimal frontier and also providing the decision-maker with a set of solutions in a region of interest to decision makers, but this approach may only be a portion of the true Pareto-optimal front, thereby reducing the number and dimensionality of target solutions, arriving at a biased distribution of Pareto-optimal solutions (Deb and Saxena 2005).
(3) Objective reduction: because of considering the importance of complete Pareto-optimal front, this approach is more suitable for solving the MaOPs with redundant objectives. The greatest difficulty of MaOPs is the curse of dimensionality (Deb et al. 2002), that is to say, if N points are needed for adequately representing a one-dimensional Pareto-optimal front, N_m points will be necessary to represent an M-dimensional Pareto-optimal front. A lower dimensional Pareto-optimal front can be possessed by eliminating objectives that are redundant.

Over the past decades, a number of multi-objective optimization algorithms have been developed to solve the multi-objective optimization problem. The techniques include non-dominated sorting genetic algorithm-II (NSGA-II) (Murugan et al. 2009; Basu 2008), multi-objective particle swarm optimizer (MOPSO) (Wang and Singh 2008; Niknam et al. 2012), multi-objective differential evolution algorithm (MODE) (Varadarajan and Swarup 2008), etc. Inspired by a multi-objective evolutionary algorithm, group search optimizer with multiple producers (GSOMP) (Guo et al. 2012), this chapter proposes a multi-objective group search optimizer with adaptive covariance and Lévy flights (MGSO-ACL) (Zheng et al. 2015) to solve the presented multi-objective optimization problem of the optimal power dispatch of an LSIES with distributed DHCs and wind power interconnected via a power grid.

The MGSO-ACL consists of three types of group members: producers, scroungers, and rangers. In each generation, the members conferred with the best fitness value of each objective are chosen as the producers, and a number of members are randomly selected as the scroungers, then the rest of members are named the rangers. The producers are assigned to search for the best fitness value for their corresponding objectives, and perform the crappie search behavior which is characterized by maximum pursuit angle, maximum pursuit distance, and maximum pursuit height (O'Brien et al. 1986). The scroungers employ the concepts based on covariance matrix adaptation evolution strategy (Hansen and Ostermeier 1996; Hansen et al. 2003) to design optimum searching strategy. Moreover, Lévy flights, which are found to be more efficient than random walks for searching resource (Viswanathan et al. 1999; Reynolds et al. 2007), are employed by the rangers to increase the diversity of group in this chapter. Applying the MGSO-ACL, a Pareto-optimal set can be obtained. The Pareto-optimal set contains all the feasible and optimal solutions, called Pareto-optimal solutions. In addition, the quality of the Pareto-optimal solutions can be measured by the metrics utilizing the index of inverted generational distance (IGD), hypervolume (HV) (Wu and Liao 2013), the mean Euclidian distance (MED), the spacing index, and the number of Pareto-optimal solutions (NPS) (Durillo et al. 2010; de Athayde Costa e Silva et al. 2013).

The optimal power dispatch of an integrated energy system consisting of distributed DHCs and wind power interconnected via a power grid is formulated as a multi-objective optimization problem mathematically. The objectives can be addressed for the economy and reliability viewpoint of both the power grid and the DHCs. Moreover, the optimization problem must satisfy various constraints aforementioned to maintain the stable operation of the LSIES. Consequently, the problem is a complex multi-objective optimization problem addressed with interval inequality constraints and nonlinear equality constraints, and it is tackled by the proposed multi-objective group search optimizer with adaptive covariance and Lévy flights (MGSO-ACL).

3.1.2 *Nonlinear Constraints Handling*

The Newton–Raphson method is widely used to solve the nonlinear power flow equations (Viana et al. 2013). However, in order to improve the computational efficiency, the fast coupled flow method (Rao et al. 1982) is applied to solve the equations.

The power flow Eqs. 3.6 and 3.7 can be rewritten in a general form: $f(V, \theta) = 0$, where V and θ are the voltage magnitude and phase angle of each bus node, respectively. The corresponding Jacobian matrix, J, is the first derivative of $f(V, \theta)$. According to the definition of admittance in power systems, the value of self-admittance is much larger than that of the injected power in a certain node (Viana et al. 2013). As a result, the Jacobian matrix of the power grid can be decoupled into

$$-\begin{bmatrix} J_H & J_N \\ J_M & J_L \end{bmatrix} \begin{bmatrix} V\Delta\theta \\ \Delta V \end{bmatrix} = \begin{bmatrix} \Delta P/V \\ \Delta Q/V \end{bmatrix} \tag{3.1}$$

where the off-diagonal elements can be neglected because the resistor of transmission line is much smaller than the reactance. Therefore, $J_N = 0$, $J_M = 0$, and J_H is set as the node admittance matrix B' while J_L is the imaginary part B'' of the node admittance matrix excluding the generator nodes. Consequently, the modified equations are $-B'\Delta\theta = \Delta P/V$ and $-B''\Delta V = \Delta Q/V$.

As a consequence, the variables (V, θ) are updated in the kth iteration as follows:

$$\begin{cases} \Delta V^{(k)} = -B''^{-1}\Delta Q(V^{(k)}, \theta^{(k)})/V^{(k)} \\ V^{(k+1)} = V^{(k)} + \Delta V^{(k)} \end{cases} \tag{3.2}$$

$$\begin{cases} \Delta\theta^{(k)} = -B'^{-1}\Delta P(V^{(k+1)}, \theta^{(k)})/V^{(k+1)} \\ \theta^{(k+1)} = \theta^{(k)} + \Delta\theta^{(k)} \end{cases} \tag{3.3}$$

By solving the power flow equations, both the control variables and dependent variables of the formulated power dispatch problem can be obtained for the next optimization iteration. Furthermore, the boundary limits also need to be tackled. As shown in the flowchart, the violation check of limits are executed during every iteration in this chapter, and once the limits are violated, the corresponding population will be dragged back into the feasible region randomly.

3.1.3 Pareto-Dominance Principle

The Pareto-dominance principle works based on the dominance concept to obtain a set of optimal non-dominated solutions called the Pareto-optimal set. The vector X_1 dominates X_2 if

$$\forall i,\ F_i(X_1) \leq F_i(X_2), \text{ and } \exists j,\ F_j(X_1) < F_j(X_2) \tag{3.4}$$

As shown in the flowchart shown in Fig. 3.1, the principle is applied by the proposed methodology to obtain the Pareto-optimal solutions during the multi-objective optimization procedure.

3.2 Multi-objective Group Search Optimizer with Adaptive Covariance and Lévy Flights

The flowchart of the proposed multi-objective group search optimizer with adaptive covariance and Lévy flights is shown in Fig. 3.1. An individual of the optimization algorithm, $\boldsymbol{x}_i$, represents the producer, scrounger, or ranger. It is a variable of the considered power dispatch problem. All the individuals form a vector consisting of all the variables. The producers, scroungers, and rangers are classified based on the fitness value of all the individuals (He et al. 2009). The detailed explanations of the steps are discussed as follows.

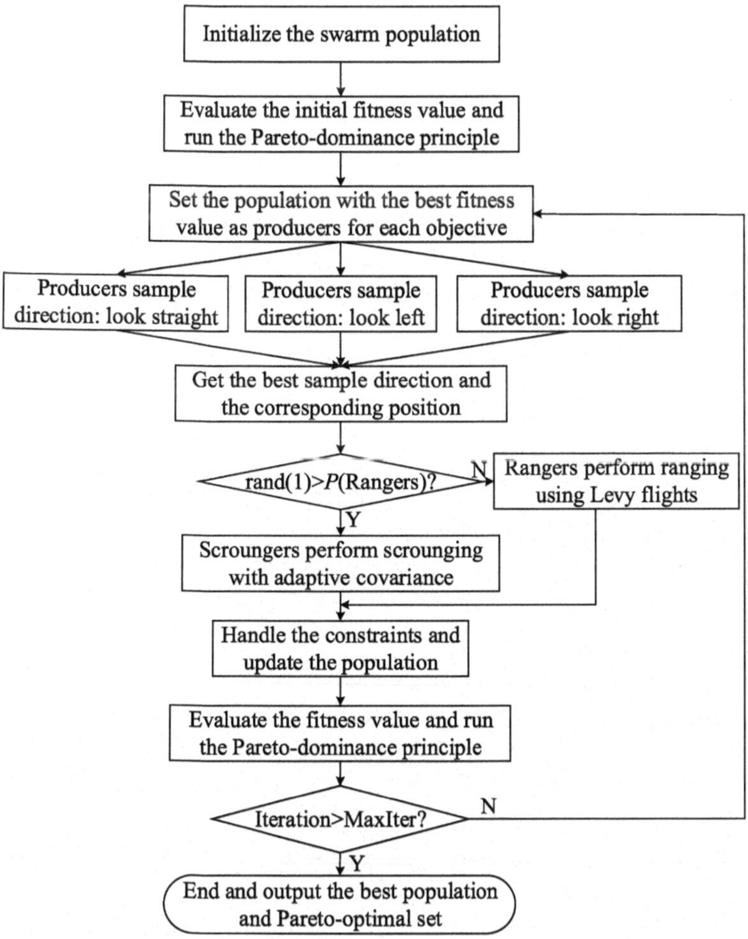

Fig. 3.1 The flowchart of the multi-objective optimization algorithm

3.2.1 Producer Searching Strategy

In MGSO-ACL, the number of producers is equal to the number of objectives (N_{ob}), which means each producer is assigned to find the best fitness value $F_p(\boldsymbol{x}_p^{(g)})$, ($p = 1, \cdots, N_{ob}$) of its corresponding objective. The searching mechanism of a certain producer is similar to that of the original GSO. Inspired by the foraging behavior of animals, it uses a scanning mechanism to randomly sample three different directions: straight, left- and right-hand side hypercube, respectively, which are formulated as follows:

$$\boldsymbol{x}_s = \boldsymbol{x}_p^{(g)} + r_1 l_{\max} \boldsymbol{D}_p^{(g)}(\varphi^{(g)}) \tag{3.5}$$

$$\boldsymbol{x}_l = \boldsymbol{x}_p^{(g)} + r_1 l_{\max} \boldsymbol{D}_p^{(g)}(\varphi^{(g)} - \boldsymbol{r}_2 \theta_{\max}/2) \tag{3.6}$$

$$\boldsymbol{x}_r = \boldsymbol{x}_p^{(g)} + r_1 l_{\max} \boldsymbol{D}_p^{(g)}(\varphi^{(g)} + \boldsymbol{r}_2 \theta_{\max}/2) \tag{3.7}$$

where $r_1 \in \mathbb{R}^1$ is a normally distributed random number with mean 0 and standard deviation 1, $\boldsymbol{r}_2 \in \mathbb{R}^{n-1}$ is a uniformly distributed random sequence in the range (0,1), $\varphi_i^{(g)} \in \mathbb{R}^{n-1}$ is the head angle and the unit vector $\boldsymbol{D}(\varphi) \in \mathbb{R}^n$ can be calculated from φ via a polar to Cartesian coordinate transformation (He et al. 2009).

If the best point has a better resource than its current position, then the producer will fly to this point. Otherwise the producer will stay in its current position and turn its head to a new randomly generated angle:

$$\varphi^{(g+1)} = \varphi^{(g)} + \boldsymbol{r}_2 \alpha_{\max} \tag{3.8}$$

where $\alpha_{\max} \in \mathbb{R}^1$ is the maximum turning angle.

On the other hand, if the producer cannot find a better area after a generations, it will turn its head back to zero degree:

$$\varphi^{(g+a)} = \varphi^{(g)} \tag{3.9}$$

where $a \in \mathbb{R}^1$ is a constant.

3.2.2 Scroungers' Behaviors with Adaptive Covariance

In this section, the adaptive covariance matrix obtained by cumulatively learning for the information organized from the group members of each generation, is employed to get a reliable estimator for determining the evolution path and step size for the scroungers' behaviors. The scroungers mainly perform the following three tasks: (1) Scroungers partition the group members into an elite group and an inferior group based on their fitness values, then the information gathered from the elite group members are used to generate a mean vector $\boldsymbol{m}$ by exponential weighting; (2) covariance matrix $\boldsymbol{C}$, which is used to obtain an estimator for determining the evolution path and step size, is updated by the mean vector; (3) the offsprings of scroungers are updated by the evolution path and step size.

The offspring of kth organizer, $\boldsymbol{x}_k^{(g+1)}$, can be modeled as follows (Auger and Hansen 2012):

$$\boldsymbol{x}_k^{(g+1)} = \boldsymbol{m}^{(g)} + \sigma^{(g)} \mathcal{N}(\mathbf{0}, \boldsymbol{C}^{(g)}) \quad k = 1, \cdots, \lambda \tag{3.10}$$

where $\mathcal{N}(\mathbf{0}, \mathbf{I})$ means a multivariate normal distribution with zero mean and unity covariance matrix, $\sigma > 0$ is the step size, λ is the number of the scroungers, superscript g denotes the generation number, $(g = 0, 1, 2, \cdots)$, and n is the dimension of the function.

Mean vector $\boldsymbol{m}^{(g)}$ of the searching distribution is a weighted average of μ successful individuals selected from the sample $\boldsymbol{x}_1^{(g)}, \cdots, \boldsymbol{x}_\lambda^{(g)}$. Covariance matrix $\boldsymbol{C}$ is updated based on mean vector, and the evolution path and step size are accordingly determined by the covariance matrix (Auger and Hansen 2012).

3.2.3 Rangers' Walks

In this chapter, Lévy flights (Yang 2010) are introduced as rangers' searching technique rather than the random walks. The step size value of the ith ranger is chosen randomly as follows:

$$s_i = 0.01 \left(\frac{u_i}{v_i} \right)^{1/\beta} (\boldsymbol{x}_i^{(g)} - \boldsymbol{x}_p^{(g)}) \tag{3.11}$$

where $u = \phi \mathrm{randn}(n)$, $v = \mathrm{randn}(n)$, $\beta = 1.50$, n is the number of variables. The $\mathrm{randn}(n)$ function generates a uniform integer between $[1, n]$, and the ϕ is computed by

$$\phi = \left(\frac{\Gamma(1+\beta)\sin(\pi\beta/2)}{\Gamma((1+\beta)/2)\beta 2^{(\beta-1)/2}} \right)^{1/\beta}$$

where Γ denotes the $gamma function$.

Consequently, rangers will move to the new point following the direction as

$$\boldsymbol{x}_i^{(g+1)} = \boldsymbol{x}_i^{(g)} + \mathrm{randn}(n) s_i \tag{3.12}$$

In this way, the individuals, $\boldsymbol{x}_i$, of the MGSO-ACL are updated according to the fitness value of the multiple objectives.

3.3 Multi-objective Group Search Optimizer with Adaptive Covariance and Chaotic Search

The MGSOACC consists of three types of group members: producers, scroungers, and rangers. In each search generation, the number of the producers is equal to that of the objectives and each producer corresponds to the member with respect to the best fitness value of the objective. The producers will scan the search field using white crappie's scanning strategies which are characterized by the maximum pursuit angle, maximum pursuit distance and maximum pursuit height (Wu et al. 2008) to seek the optimal resource. The scroungers adopt the adaptive covariance matrix (Hansen et al. 2003) in order to make the search strategy of scroungers be adaptive and to get a reliable estimator for the paths and thus could enhance the local searchability of the proposed algorithm. The detailed introduction to producers and scroungers can be found in Wu et al. (2008).

In this section, chaotic search is employed as the rangers' search strategy to maintain the diversity of the group (Strogatz 2014). Chaos is a typical nonlinear phenomenon in nature which is characterized by ergodicity, randomicity and sensitivity to its initial conditions (Strogatz 2014). Because of the ergodicity and randomicity, chaotic search is often incorporated into other evolutionary algorithms to enhance their searchability (Jia et al. 2011; Talatahari et al. 2012). First, the chaotic sequence is generated based on the logistic map (Strogatz 2014)

Table 3.1 Peseudocode of multi-objective interval optimization using MGSOACC

```
Set g := 1;
Input the parameters of the integrated energy system;
Initialize parameters of each member of MGSOACC;
Input the prediction interval of wind speed and solar irradiation;
Obtain the lower and upper bounds of the objective interval of each member by non-linear programming using (5.32);
Calculate the fitness values of initial members using (5.33);
WHILE (the termination conditions are not met)
    FOR (each member in the group)
        Choose producers :    Select producers from the group. The number of producers is equal to the number
                              of objectives. The member with the best fitness value of the pth objective is selec-
                              ted as producer;
        Perform producing :   Each producer scans at zero degree and then scan laterally by randomly sampling
                                  three points in the scanning field using (5)-(9) in [231];
        Perform scrounging :  Except the producers, randomly select 70% from the rest members to perform scr-
                                  ounging:
                                  1) Generate mean vector by exponential weighting [71];
                                  2) Update covariance matrix to determine evolution path and update step-size us-
                                  ing (30) in [256];
        Perform ranging :     Except the producers and scroungers, the rest members perfom ranging:
                                  1) Generate the chaotic sequence using (6.26);
                                  2) Rangers perform chaotic search using (6.27);
        Update group :        Select new producers and generate new group members;
    END FOR
    Calculate fitness :       1) Obtain the lower and upper bounds of the objective interval of each current me-
                                  mber by non-linear programming using (5.32);
                                  2) Calculate the fitness values of current members using (5.33);
    Pareto selection :        Update the Pareto solutions using fast non-dominated sorting technology and fix
                                  the number of elements in the Pareto solution set as a constant by the crowded-
                                  comparison operator [36];
    g = g + 1;
END WHILE
```

$$u^{(g+1)} = \mu \cdot u^{(g)} \cdot (1 - u^{(g)}) \tag{3.13}$$

where $\mu = 4$ is the control parameter, $u^{(g)} \in [0, 1] \wedge u^{(0)} \notin \{0.0, 0.25, 0.50, 0.75, 1.0\}$. g denotes the gth iteration.

After that, the position of the ith ranger is updated based on the chaotic search shown as follows:

$$\boldsymbol{x}_i^{(g+1)} = \boldsymbol{x}_i^{(g)} + u^{(g+1)} \cdot (\boldsymbol{x}_i^{(g)} - \mathscr{X}_i). \tag{3.14}$$

where $\mathscr{X}_i$ is the Pareto-optimal solution selected from Pareto-set randomly. The peseudocode of the multi-objective interval optimization using MGSOACC is given in Table 3.1.

3.4 Multi-objective Evolutionary Predator and Prey Strategy

The EPPS is a population-based optimization algorithm, which takes inspiration from the group living behaviors of dingo hunting and sheep escaping. The population of EPPS is called a group and each individual within the group is called a member. Each member represents a position vector of a n−dimensional search space and is randomly positioned at the beginning. Here, n is the dimension of the objective function. In each generation, the members of the group are classified into four different types, representing experienced predators, strategic predators, the prey, and its safe location, to cope with three typical scenarios, predators hunting, prey scanning, and prey escaping.

In EPPS, the member that corresponds to the best fitness value of the group is chosen as the prey, and the member that corresponds to the worst fitness value of the group is chosen as the safe location of the prey; the rest of the members are classified randomly as either the experienced predator or the strategic predator. The EPPS investigates three processes from the perspective of the predators' hunting and the prey's escaping. When a group of predators locks onto a prey, the experienced predators run experientially for hunting; the prey realizes the danger and tries to escape from its dilemma by scanning for its safe location; as for the strategic predators, they run strategically for hunting. The searching behaviors of the predators and the prey are described in detail as follows.

3.4.1 *Experienced Predators' Searching Mechanism*

For each search generation, a number of group members are selected as the experienced predators. The experienced predators will determine their search paths by accumulatively learning for the successful paths of the predators of the group. Here, the successful paths indicate the directions of fitness value decreasing. In order to get a reliable estimate for the paths, the experienced predators adopt the concept of adaptive covariance matrix (Hansen et al. 2003). The adaptive mechanism is based on the assumption that the successful evolutionary paths of the predators used in recent past generations may also be successful in the following generation. Gradually, the most suitable evolutionary paths can be developed automatically to guide the search behavior of each experienced predator in different evolutionary stages. The predatory behavior of the ith experienced predator at generation $(g + 1)$ can be modeled as follows:

$$\boldsymbol{x}_i^{(g+1)} = \boldsymbol{m}^{(g)} + \sigma^{(g)} \mathcal{N}(\mathbf{0}, \boldsymbol{C}^{(g)}), \quad i = 1, \cdots, \lambda \tag{3.15}$$

where $\boldsymbol{m}$ and $\boldsymbol{C}$ are mean value and covariance matrix of the predators, respectively, developed by the position vectors of the predators of the group, $\mathcal{N}(\mathbf{0}, \mathbf{I})$ corresponds to a multivariate normal distribution with zero mean and unity covariance matrix, $\sigma(\sigma > 0)$ is the step size and λ is the number of the experienced predators.

During the search process of EPPS, if an experienced predator finds a better location than the current prey and other predators, in the next search generation, it will switch to be the prey and all the other predators, including the prey in the previous search generation, will perform a hunting mechanism; and if an experienced predator finds a worse location than the current safe location, in the next search generation, it will switch to be the safe location and the prey performs the escaping mechanism to this location. The prey and the strategic predator, which will be introduced in the following paragraphs, are also implementing these switching mechanisms in each search process. Thus, different types of members can play different roles during each search generation, and even the same member can play different roles during different search generations. Thereupon, EPPS could escape from local minima in the earlier search bouts and obtain a good balance between its local exploration and global exploitation abilities.

3.4.2 *Prey's Searching Mechanism*

During each search generation, $\boldsymbol{x}_p$ and $\boldsymbol{x}_s$ denote the prey and its safe location, respectively. When the prey is aware of its dangerous situation, it will scan its safe location so as to escape from the dilemma.

In order to obtain an efficient search performance, the basic scanning strategy inspired from white crappies (O'Brien et al. 1986), which is characterized by maximum pursuit angle and maximum pursuit height, is employed as the scanning directions of the prey. Additionally, a scanning distance shown in Eq. 3.5 has been proposed based on the position vectors of the prey and its safe location to shorten the prey's scope for searching, which enables the prey to explore the unreachable areas with a higher probability than that reached by scanning the whole search scope (He et al. 2009).

Based on the scanning directions and scanning distance, the prey initially scans at zero degree by using Eq. (3.16), and then scans laterally by randomly sampling two points in the scanning field by using Eqs. (3.17) and (3.18). At the $(g+1)$th generation, the first position that the prey escapes by scanning at zero degree

$$\boldsymbol{x}_z = \boldsymbol{x}_p^{(g)} + l_{\max}^{(g)} \boldsymbol{D}_p^{(g)}(\varphi^{(g)}) \tag{3.16}$$

the second position in the right-hand side

$$\boldsymbol{x}_r = \boldsymbol{x}_p^{(g)} + l_{\max}^{(g)} \boldsymbol{D}_p^{(g)}(\varphi^{(g)} + \boldsymbol{r}_1 \theta_{\max}/2) \tag{3.17}$$

and the third position in the left-hand side

$$\boldsymbol{x}_l = \boldsymbol{x}_p^{(g)} + l_{\max}^{(g)} \boldsymbol{D}_p^{(g)}(\varphi^{(g)} - \boldsymbol{r}_1 \theta_{\max}/2) \tag{3.18}$$

where $\boldsymbol{r}_1 \in \mathbb{R}^{n-1}$ is a uniformly distributed random sequence in the range (0,1), $\varphi^{(g)} \in \mathbb{R}^{n-1}$ is the heading angle and the unit vector $\boldsymbol{D}(\varphi) \in \mathbb{R}^n$ can be calculated from φ via a polar to Cartesian coordinate transformation (Mustard 1964).

The scanning distance at the gth generation can be calculated as follows:

$$l_{\max}^{(g)} = ||\boldsymbol{x}_p^{(g)} - \boldsymbol{x}_s^{(g)}|| = \sqrt{\sum_{i=1}^{n} (x_{p_i}^{(g)} - x_{s_i}^{(g)})^2} \tag{3.19}$$

where x_{p_i} and x_{s_i} are the elements for the ith dimension of $\boldsymbol{x}_p$ and $\boldsymbol{x}_s$, respectively.

If the fitness value of the current prey is worse than one of the other members' fitness values, in the following generation, the new prey will be chosen from the group and execute an escaping mechanism by turning its head to a new randomly generated angle

$$\varphi^{(g+1)} = \varphi^{(g)} + \boldsymbol{r}_1 \alpha_{\max} \tag{3.20}$$

where $\alpha_{\max} \in \mathbb{R}^1$ is the maximum turning angle.

If the fitness value of the current prey does not change after a generations, it would turn its head back to zero degree

$$\varphi^{(g+a)} = \varphi^{(g)} \tag{3.21}$$

where $a \in \mathbb{R}^1$ is a constant.

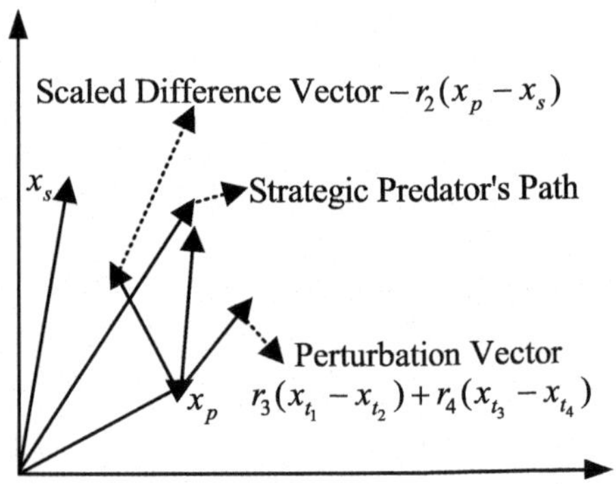

Fig. 3.2 The typical search path of the strategic predator

3.4.3 Strategic Predators' Searching Mechanism

The remaining members are selected as the strategic predators. Compared to the experienced predators, the strategic predators will adjust their search paths according to the prey's position in each search bout. That is, the strategic predators do not run to the prey's position directly. Instead, they will flock to the prey's escaping direction, which is developed based on the position vectors of the prey and its safe location. At the $(g+1)$th generation, the ith strategic predator can be described as

$$\boldsymbol{x}_j^{(g+1)} = \boldsymbol{x}_p^{(g)} - r_2 \cdot (\boldsymbol{x}_p^{(g)} - \boldsymbol{x}_s^{(g)}), \quad j = 1, \cdots, \mu \tag{3.22}$$

However, in reality, the strategic predators may not be able to exactly catch the prey's escaping traces. In consideration of this, a perturbation vector has been added to Eq. (3.22) so as to maintain group diversity for jumping out of the potential local optima:

$$\begin{aligned} \boldsymbol{x}_j^{(g+1)} &= \boldsymbol{x}_p^{(g)} - r_2 \cdot (\boldsymbol{x}_p^{(g)} - \boldsymbol{x}_s^{(g)}) \\ &+ r_3 \cdot (\boldsymbol{x}_{t_1}^{(g)} - \boldsymbol{x}_{t_2}^{(g)}) + r_4 \cdot (\boldsymbol{x}_{t_3}^{(g)} - \boldsymbol{x}_{t_4}^{(g)}) \end{aligned} \tag{3.23}$$

where $\boldsymbol{x}_{t_1} \neq \boldsymbol{x}_{t_2} \neq \boldsymbol{x}_{t_3} \neq \boldsymbol{x}_{t_4}$ are randomly chosen from the set of strategic predators, r_2, r_3, r_4 are the uniformly distributed random numbers in the range [0, 1], and μ is the number of the strategic predators.

Equation (3.23) can help the strategic predators explore more resources distributed around the prey and achieve their own historical best positions, rather than crowd around the prey that is likely to be associated with a local optimum. The strategic predator's search path is shown in Fig. 3.2.

In order to maximize the chances of finding resources, there are several strategies to restrict their search to a profitable patch. One of the most efficient strategies is turning back into a patch when its edge is detected (Dixon 1959). In this section, EPPS employs this strategy to handle the bounded search space: when a member

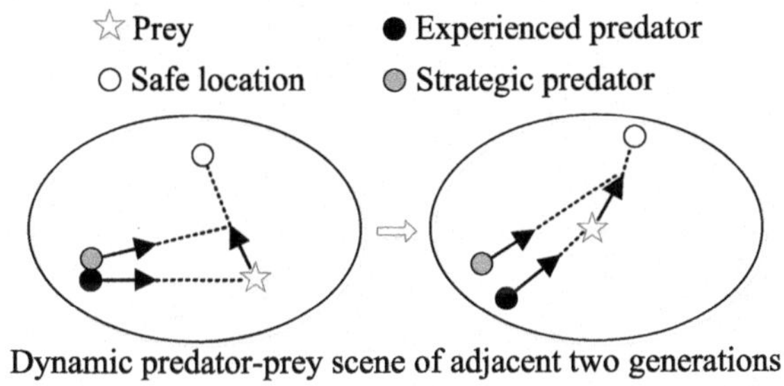

Fig. 3.3 The typical search paths of the predators and the prey

is outside the search space, it will turn back into the search space by setting the variables that violated bounds to its previous values.

It can be seen from the above description that the searching mechanisms of EPPS are similar to DE, PSO, or covariance matrix adaptative evolution strategy (CMA-ES) (Hansen et al. 2003). However, there are many differences with the most notable being the distinction in the concept. EPPS is inspired from animal searching behavior and group living theory which are adopted to develop an escaping mechanism and a classification mechanism to construct a good balance between its local search and global search abilities. Our EPPS consists of experienced predators, strategic predators, the prey and its safe location, which have never been used to develop an evolutionary computation algorithm. In addition, the search mechanisms of EPPS are radically different from those of DE and PSO. The predator–prey procedure of adjacent two generations in EPPS is presented in Fig. 3.3. It is worth mentioning that in order to depict a dynamic predator–prey scenario, in this figure, we artificially placed the prey and its safe location into two different positions, respectively, in the adjacent two generations. Therefore, the experienced predators and the strategic predators will adjust their search directions, respectively. The steps involved are presented in Algorithm 1.

3.4.4 Numerical Studies

3.4.4.1 Experiments Setting

To evaluate the applicability of EPPS, we carry out numeral experiments on 20 canonical benchmark functions in 30-dimensional case. These test functions, which are shown in Table 3.2, can be classified into three groups. The first seven functions $f_1 - f_7$ are unimodal functions. As the preservation of diversity in many EAs is at the cost of slower convergence, the unimodal functions are used to test if EPPS has the feature of fast convergence. The next seven functions $f_8 - f_{14}$ are multimodal functions with many local optima. These functions are used to test the global search-ability of EPPS in avoiding premature convergence. Finally, the last six functions $f_{15} - f_{20}$ are shifted and rotated with more complex characteristics, and can be used

Generate initial group and set up parameters for each member;
Evaluate each member, and determine the prey and the safe location;
Set $g := 0$, the maximum number of generations $:= Max\_Gen$, population size $:= Pop$, and the index of the prey $:= index^{(0)}$;
while $g \leq Max\_Gen$ **do**
  Calculate the scanning distance, $l_{\max}^{(g)}$, according to Eq. 3.5;
  for $i = 1 : Pop$ **do**
    if $i == index^{(g)}$ **then**
      Perform prey's search mechanism according to (2)-(4);
    else if $rand < 0.3$ **then**
      Perform experienced predators' search mechanism according to (1);
    else
      Perform strategic predators' search mechanism according to (9);
    end if
  end for
  Modify the position of each members to satisfy the constraints, if necessary;
  Calculate the fitness values of each member, and update the prey, the safe location and $index$;
  $g = g + 1$;
end while

Algorithm 1: The pseudocode of EPPS algorithm

to compare the performance of different algorithms in a more systematic manner (Chen et al. 2013).

All the simulations are carried out utilizing MATLAB 7.11 on an Intel Core i5, 3.1 GHz computer with 4 GB RAM. During each run, the maximum number of function evaluations is set to 150,000 for $f_1 - f_{14}$ and 300,000 for $f_{15} - f_{20}$, respectively. In order to make a coherent comparison, the fitness values below 10^{-16} are assumed to be 0 in all experiments. To validate the effectiveness of EPPS, we compared EPPS with group search optimizer (GSO) (He et al. 2009), the latest standard PSO (SPSO) (Omran and Clerc 2011), covariance matrix adaptative evolution strategy (CMA-ES) (Hansen et al. 2003), and differential evolution (DE) (Storn and Price 1997). The selection of these EAs in the comparison is based on the following reasons. For one thing, the CMA-ES and SPSO are characterized by its fast-converging feature on simple unimodal functions. Therefore, by comparing EPPS with CMA-ES and SPSO, we can learn whether EPPS can present the fast-converging feature. For another, GSO and DE selected in the comparisons are representative and well-performed algorithms in terms of global searchability. Thus, by comparing EPPS with GSO and DE, we can learn whether EPPS can prevent premature convergence while still maintain the fast-convergence feature as well. To reduce statistical errors, each test is repeated for 30 times independently.

The parameters of EPPS are set as follows: $a = round(\sqrt{n+1})$ ($round(X)$ rounds the elements of X to the nearest integers), $\theta_{\max} = \pi/a^2$, $\alpha_{\max} = \theta_{\max}/2$, the population size is 200, $\sigma^{(0)} = 0.5$, and the percentage of the strategic predators is 30%. These parameters settings are, empirically, applied for all the benchmark functions used in this section. Parameter settings of the GSO, DE, CMA-ES and

Table 3.2 Twenty high-dimensional benchmark functions, where n is the dimension of the function, S is the search range, and $f_{\min}$ is the global minimum value of the function

Unimodal functions	n	S	$f_{\min}$
$f_1 = \sum_{i=1}^{n} x_i^2$	30	$[-100, 100]^n$	0
$f_2 = \sum_{i=1}^{n} \lvert x_i \rvert + \prod_{i=1}^{n} \lvert x_i \rvert$	30	$[-10, 10]^n$	0
$f_3 = \sum_{i=1}^{n} \left(\sum_{j=1}^{i} x_j\right)^2$	30	$[-100, 100]^n$	0
$f_4 = (x_1 - 1)^2 + \sum_{i=2}^{n} i(2x_i^2 - x_{i-1})^2$	30	$[-10, 10]^n$	0
$f_5 = \sum_{i=1}^{n} (\lfloor x_i + 0.5 \rfloor)^2$	30	$[-100, 100]^n$	0
$f_6 = \sum_{i=1}^{n} ix_i^4 + random[0, 1)$	30	$[-1.28, 1.28]^n$	0
$f_7 = \sum_{i=1}^{n} ix_i^2$	30	$[-10, 10]^n$	0
Multimodal functions	n	S	$f_{\min}$
$f_8 = \sum_{i=1}^{n-1} (100(x_i^2 - x_{i+1})^2 + (x_i - 1)^2)$	30	$[-30, 30]^n$	0
$f_9 = -\sum_{i=1}^{n} (x_i \sin(\sqrt{\lvert x_i \rvert}))$	30	$[-500, 500]^n$	-12569.5
$f_{10} = \sum_{i=1}^{n} (x_i^2 - 10\cos(2\pi x_i) + 10)^2$	30	$[-5.12, 5.12]^n$	0
$f_{11} = -20\exp\left(-0.2\sqrt{\frac{1}{n}\sum_{i=1}^{n} x_i^2}\right) - \exp(\frac{1}{n}\sum_{i=1}^{n}\cos(2\pi x_i)) +$ $20 + e$	30	$[-32, 32]^n$	0
$f_{12} = \frac{1}{4000}\sum_{i=1}^{n} x_i^2 - \prod_{i=1}^{n}\cos(\frac{x_i}{\sqrt{i}}) + 1$	30	$[-600, 600]^n$	0
$f_{13} = \frac{\pi}{n}\left\{10\sin^2(\beta y_1) + \sum_{i=1}^{29} (y_i - 1)^2[1 + 10\sin^2(\beta y_{i+1})]\right.$ $\left.+(y_n - 1)^2\right\} + \sum_{i=1}^{30} u(x_i, 10, 100, 4)$ $y_i = 1 + \frac{1}{4}(x_i + 1)$	30	$[-50, 50]^n$	0
$f_{14} = 0.1\left\{\sin^2(\pi 3x_1) + \sum_{i=1}^{29} (x_i - 1)^2[1 + \sin^2(3\pi x_{i+1})]\right.$ $\left.+(x_n - 1)^2[1 + \sin^2(2\pi x_{30})]\right\} + \sum_{i=1}^{30} u(x_i, 5, 100, 4)$	30	$[-50, 50]^n$	0
Rotated and shifted multimodal functions	n	S	$f_{\min}$
$f_{15} = \sum_{i=1}^{n-1} (100(z_i^2 - z_{i+1})^2 + (z_i - 1)^2)$ $\boldsymbol{z} = \boldsymbol{M}(\frac{2.048(\boldsymbol{x}-\boldsymbol{o})}{100} + 1)$	30	$[-100, 100]^n$	0
$f_{16} = -20\exp\left(-0.2\sqrt{\frac{1}{n}\sum_{i=1}^{n} z_i^2}\right) - \exp(\frac{1}{n}\sum_{i=1}^{n}\cos(2\pi z_i))$ $+20 + e, \boldsymbol{z} = \boldsymbol{M}((\boldsymbol{x} - \boldsymbol{o})$	30	$[-100, 100]^n$	0
$f_{17} = \sum_{i=1}^{n} (\sum_{k=0}^{k\max} [a^k \cos(2\pi b^k (z_i + 0.5))]) -$ $n\sum_{k=0}^{k\max} [a^k \cos(2\pi b^k \cdot 0.5)]$ $a = 0.5, b = 3, k\max = 20, \boldsymbol{z} = \boldsymbol{M}(\frac{0.5(\boldsymbol{x}-\boldsymbol{o})}{100})$	30	$[-100, 100]^n$	0
$f_{18} = \frac{1}{4000}\sum_{i=1}^{n} z_i^2 - \prod_{i=1}^{n}\cos(\frac{z_i}{\sqrt{i}}) + 1, \boldsymbol{z} = \boldsymbol{M}(\frac{600(\boldsymbol{x}-\boldsymbol{o})}{100} + 1)$	30	$[-100, 100]^n$	0
$f_{19} = \frac{10}{n^2}\prod_{i=1}^{n} (1 + i\sum_{j=1}^{32} \frac{\lvert 2^j z_i - round(2^j z_i) \rvert}{2^j})^{\frac{10}{n^{1.2}}} - \frac{10}{n^2}$ $\boldsymbol{z} = \boldsymbol{M}(\frac{5(\boldsymbol{x}-\boldsymbol{o})}{100} + 1)$	30	$[-100, 100]^n$	0
$f_{20} = g(z_1, z_2) + g(z_2, z_3) + \cdots + g(z_{n-1}, z_n) + g(z_n, z_1)$ $g(x, y) = 0.5 + \frac{(\sin^2(\sqrt{x^2+y^2}) - 0.5)}{(1+0.001(x^2+y^2))^2}, \boldsymbol{z} = \boldsymbol{M}(\boldsymbol{x} - \boldsymbol{o}) + 1$	30	$[-100, 100]^n$	0

*In f_{13} and f_{14}, $u(x_i, k_1, k_2, k_3) = \begin{cases} k_2(x_i - k_1)^{k_3}, & x_i > k_1 \\ 0, & -k_1 \le x_i \le k_1 \\ k_2(-x_i - k_1)^{k_3}, & x_i < -k_1 \end{cases}$.

*In $f_{15} - f_{20}$, $\boldsymbol{o}$ is a shifted vector and $\boldsymbol{M}$ is a transformation matrix. Please refer to Liao and Stutzle (2013)

SPSO used in the comparisons can be found in the original papers (He et al. 2009; Storn and Price 1997; Hansen et al. 2003; Omran and Clerc 2011), respectively.

3.4.4.2 Results Analysis

Tables 3.3, 3.4 and 3.5 lists the mean and standard deviation of the fitness values obtained by EPPS, CMA-ES, DE, GSO, and SPSO over 30 independent runs on functions $f_1 - f_{20}$. It should be mentioned that the algorithm that performs best in one problem will be highlighted in boldface. In order to assess whether the results obtained by EPPS are statistically different from the results obtained by the other four algorithms, the nonparametric statistical test called Wilcoxon Signed-Rank Tests (Conover and Conover 1980; Derrac et al. 2011) are employed for pairwise comparisons where the confidence level has been fixed to 95%. In the following tables, an h value of one indicates that the performances of the two algorithms are statistically different with 95% certainty, whereas a h value of zero implies that the performances are not statistically different. In addition, #+, #−, and # $\sim$ mean that the performance of EPPS is significantly better than, significantly worse than, and statistically equivalent to the performance of its rival in terms of the statistically test results, respectively. For example, the simulation results obtained by comparing EPPS with CMA-ES on unimodal benchmark functions are (2, 0, 5), which means that EPPS achieves significantly better results than, significantly worse results than, and statistically equivalent results to CMA-ES on 2, 0, and 5 problems, respectively. In Zhan et al. (2009), it is claimed that the convergence speed can be measured by the mean number of function evaluations required. Therefore, in the following tables, if algorithm A can obtain a smaller fitness value than algorithm B within the same number of function evaluations, it indicates that the convergence speed of algorithm A is faster than that of algorithm B.

On unimodal functions $f_1 - f_7$, it is relatively easy to converge the global optimum, and thus we focus on comparing the performance of the algorithms in terms of solution accuracy and convergence speed. From the comparison of the results on these functions, we can see that EPPS performs better than GSO in terms of the mean and the standard deviation on $f_1 - f_7$. EPPS surpasses all other algorithms on functions f_4 and f_6, and has the same performance as CMA-ES, DE and SPSO on functions f_1, f_2, and f_7. As for functions f_3 and f_5, EPPS performs better than three other algorithms on f_3 and two algorithms on f_5. According to the results of the nonparametric Wilcoxon Signed-Rank Tests, EPPS significantly outperforms four other algorithms on f_4 and f_6. Overall, EPPS manages to find accurate solutions within the same running conditions on all of these unimodal functions.

On multimodal functions $f_8 - f_{14}$, the global optimum is much more difficult to locate. Therefore, in the comparison, we can study the performance of the algorithms in terms of the solution accuracy, convergence speed, and reliability. In Table 3.5, it can be clearly seen that EPPS performs better than four other algorithms on f_8, $f_{10} - f_{14}$ and three algorithms on $f_8 - f_{14}$ in terms of mean value and standard deviation. EPPS performs the same performance with GSO on function f_9 in

Table 3.3 Comparison of EPPS with CMA-ES, DE, GSO, and SPSO on benchmark functions $f_1 - f_7$. All results have been averaged over 30 runs

f	Algorithms	EPPS	CMA-ES	DE	GSO	SPSO
1	Mean	**0**	**0**	**0**	1.9481E-8	**0**
	Std.	**0**	**0**	**0**	1.1629E-8	**0**
	h	–	0	0	1	0
2	Mean	**0**	**0**	**0**	3.7039E-5	**0**
	Std.	**0**	**0**	**0**	8.6185E-5	**0**
	h	–	0	0	1	0
3	Mean	**0**	**0**	8.7670E-6	5.7829	6.1780E-10
	Std.	**0**	**0**	1.3560E-6	3.6813	7.5592E-10
	h	–	0	1	1	1
4	Mean	**0**	0.6667	0.8664	0.1078	0.7125
	Std.	**0**	**0**	5.4937E-3	3.9981E-2	2.5918E-4
	h	–	1	1	1	1
5	Mean	**0**	**0**	**0**	1.6000E-2	0.9667
	Std.	**0**	**0**	**0**	0.1333	1.2726
	h	–	0	0	1	1
6	Mean	**1.1069E-5**	0.2180	1.0816E-2	7.3773E-2	3.6191E-3
	Std.	**1.1103E-5**	0.1692	1.1105E-2	9.2557E-2	4.8209E-2
	h	–	1	1	1	1
7	Mean	**0**	**0**	**0**	3.8926E-8	**0**
	Std.	**0**	**0**	**0**	6.7362E-8	**0**
	h	–	0	0	1	0
(#+, #−, # ∼)		–	(2,0,5)	(3,0,4)	(7,0,0)	(4,0,3)

terms of mean value. According to the results of the nonparametric Wilcoxon Signed-Rank Tests, EPPS significantly outperforms four other algorithms on f_8, $f_{10} - f_{14}$. These comparison results validate the capability of EPPS in optimizing multimodal functions in terms of solution accuracy, convergence speed, and robustness.

On shifted and rotated functions $f_{15} - f_{20}$, the dimensions of these functions become nonseparable, and thus the resulting problems become more difficult for EAs to solve. The comparison results on shifted and rotated functions are tabulated in Table 3.5. From the table, it can be seen that EPPS achieves significantly better results than four other algorithms on functions f_{15}, $f_{17} - f_{20}$, and three other algorithms on functions $f_{15} - f_{20}$. Therefore, EPPS is among one of the best performance EAs for solving these shifted and rotated functions within the compared set of algorithms.

The comparison of convergence rates among EPPS, CMA-ES, DE, GSO, and SPSO is also carried out on the seven multimodal functions, by observing the evolution of the fitness values recorded in the optimization process. Figure 3.4 only shows the convergence rates on functions $f_8 - f_{13}$ since the convergence rates on functions f_{13} and f_{14} are similar. For functions $f_{10} - f_{13}$, it is obviously that EPPS

Table 3.4 Comparison of EPPS with CMA-ES, DE, GSO, and SPSO on benchmark functions $f_8 - f_{14}$. All results have been averaged over 30 runs

f	Algorithms	EPPS	CMA-ES	DE	GSO	SPSO
8	Mean	**9.1667E-4**	29.5072	8.3493	49.8359	33.1737
	Std.	**4.2813E-3**	5.4334	2.2737	30.1771	31.2810
	h	–	1	1	1	1
9	Mean	**–12569.4882**	–8930.7622	–6680.9926	**–12569.4882**	–9776.3545
	Std.	**1.9249E-4**	672.6874	194.9523	2.2140E-2	375.5870
	h	–	1	1	0	1
10	Mean	**0**	49.2348	99.6636	2.7415	22.5064
	Std.	**0**	11.6365	12.4016	1.4651	6.3818
	h	–	1	1	1	1
11	Mean	**8.8818E-16**	14.1209	7.3477E-8	2.6548E-5	0.9879
	Std.	**0**	6.3862	1.4371E-8	3.0820E-5	0.8246
	h	–	1	1	1	1
12	Mean	**0**	6.3818E-4	1.9362E-8	3.0792E-2	5.8107E-3
	Std.	**0**	7.5655E-4	5.3913E-8	3.0867E-2	7.7213E-3
	h	–	1	1	1	1
13	Mean	**0**	3.4606E-3	1.3629E-10	2.7648E-11	1.7316E-2
	Std.	**0**	1.8862	3.3128E-10	9.1674E-11	6.1435E-2
	h	–	1	1	1	1
14	Mean	**0**	7.3223E-4	3.9023E-10	4.6948E-5	2.9116E-4
	Std.	**0**	2.7904E-3	5.0326E-10	7.0010E-4	4.9549E-4
	h	–	1	1	1	1
(#+, #−, # ∼)		–	(7,0,0)	(7,0,0)	(6,0,1)	(7,0,0)

converges much faster than the other four algorithms. As for function f_8, EPPS has a slower convergence rate at beginning and shows a faster convergence rate at end. In addition, EPPS has almost the same performance as GSO on function f_9.

3.4.4.3 Computation Complexity

In order to investigate the relationship between the dimensionality of the multimodal functions to be solved and the number of consumed function evaluations, EPPS, CMA-ES, DE, GSO, and SPSO are used to solve f_{14}, over 30 independent runs, whose dimensionality n is set to 15, 30, 50, 100, 150, 200, 250, and 300, respectively. For different dimensions of the function, the iteration will be terminated when the fitness value reaches an acceptable accuracy 1×10^{-3} or the function evaluations reach the maximum number of function evaluations 3×10^6. The reason for choosing function f_{14} is that this benchmark function is representative in function optimization

Table 3.5 Comparison of EPPS with CMA-ES, DE, GSO, and SPSO on benchmark functions $f_{15} - f_{20}$. All results have been averaged over 30 runs

f	Algorithms	EPPS	CMA-ES	DE	GSO	SPSO
15	Mean	**0**	105.1028	177.8154	45.7762	201.1326
	Std.	**0**	36.7813	58.9623	33.2185	97.6771
	h	–	1	1	1	1
16	Mean	**20.0006**	21.1835	21.0342	20.0362	20.7861
	Std.	**2.0012E-4**	0.6389	0.8325	9.6251E-2	0.2711
	h	–	1	1	0	1
17	Mean	**0.9014**	46.2053	27.1266	17.6507	28.6448
	Std.	**0.7817**	4.9311	3.6881	8.6319	2.9650
	h	–	1	1	1	1
18	Mean	**0**	13.1172	25.1039	1.4128E-3	13.6717
	Std.	**0**	11.6325	16.7864	2.6925E-3	10.1201
	h	–	1	1	1	1
19	Mean	**0.2827**	5.7714	2.3029	0.3225	1.6543
	Std.	**0.3108**	2.0183	1.4388	0.4128	1.3192
	h	–	1	1	1	1
20	Mean	**11.4117**	14.9326	13.4917	12.9826	13.2468
	Std.	**1.2309**	2.6368	2.5771	2.0117	7.8195
	h	–	1	1	1	1
(#+, #−, # ∼)		–	(5,0,0)	(5,0,0)	(4,0,1)	(5,0,0)

and the amount of the local optima of the benchmark function increases with increasing dimension. Moreover, CMA-ES, DE, GSO, and SPSO could reach the acceptable accuracy (1×10^{-3}) in the 30-dimensional case. Figure 3.5 illustrates the number of function evaluations consumed in optimizing f_{14} by the five algorithms where the dimensionality is increased from 15 to 300. From the figure, it can be clearly seen the number of function evaluations consumed by CMA-ES and SPSO increases sharply as the dimensionality linearly increases. Even worse, these two algorithms could not converge when the dimensionality reaches 50 and 60, respectively. DE and GSO could converge to the acceptable accuracy and the number of function evaluations consumed increases almost following $55000 \times e^{(n/80)}$. As for our proposed EPPS, it always converges with different dimensions and the number of function evaluations consumed increases almost following $10000 \times e^{(n/80)}$. That is to say, EPPS offers a 5.5 times higher speed than DE and GSO in optimizing f_{14}, which is measured by the mean number of function evaluations needed to reach an acceptable solution. Therefore, EPPS has more robust and much faster than CMA-ES, DE, GSO, and SPSO.

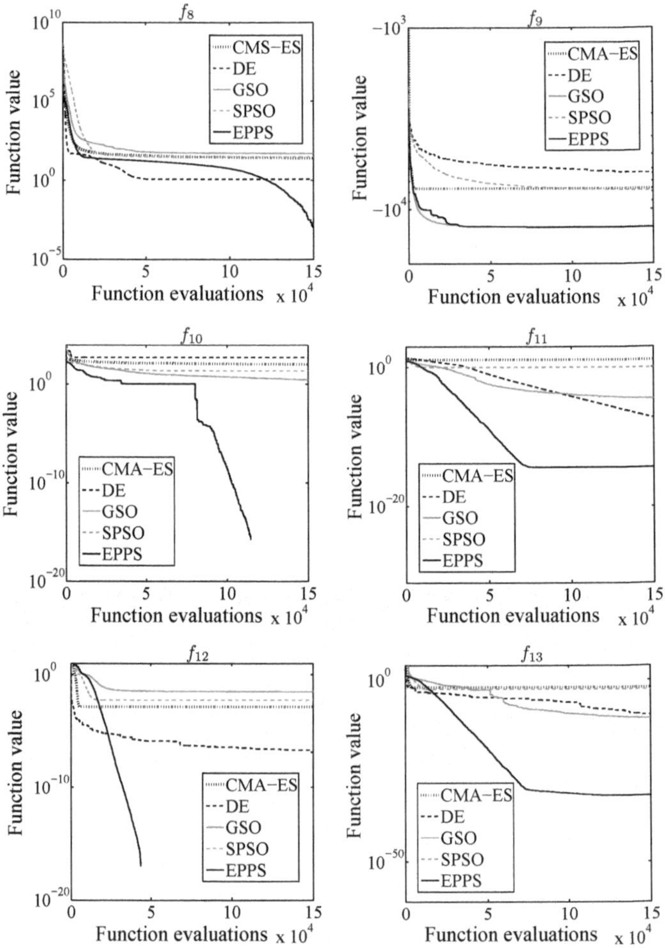

Fig. 3.4 The comparison of convergence rates among CMA-ES, DE, GSO, SPSO, and EPPS on seven multimodal benchmark functions, $f_8 - f_{13}$, respectively

3.4.4.4 Performance Analysis

In order to exhibit the optimization process of EPPS, the scanning distance, l_{max}, evaluated on functions f_8 and f_{14} has been shown in Fig. 3.6. From the figure, we can see that the scanning distance, l_{max}, does not shrink to zero with the increase of generation numbers. This indicates that EPPS could maintain a global searchability all the time with a certain level of population diversity. According to the results shown in Tables 3.3, 3.4, and 3.5, EPPS could effectively optimize these benchmark functions and obtain more accuracy solutions in most cases. This implies that EPPS achieves a good balance between its local searchability and global searchability.

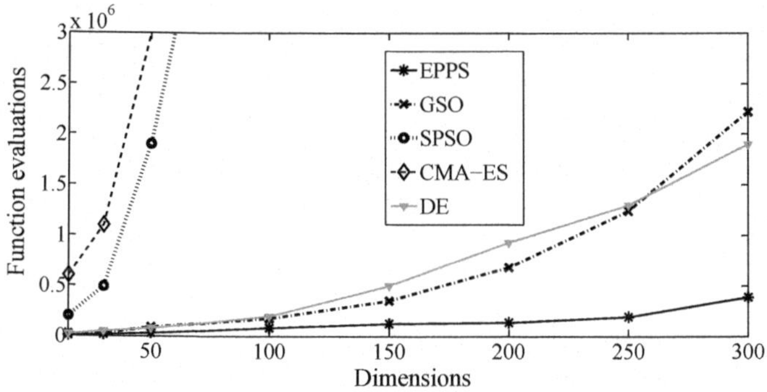

Fig. 3.5 The comparison of function evaluations consumed by EPPS, CMA-ES, DE, GSO, and SPSO with different dimensionality, on function f_{14}

According to the above comparisons and discussions, we can obtain the following conclusions. First, based on the prey's search mechanism, the diversity of the group members has been maintained in the process of the whole optimization. Thus, EPPS offers a novel global searchability. Second, the experienced predators' search mechanism can provide a reliable estimator for the evolution path and step size and thus EPPS provides an efficient local search ability. Finally, based on the strategic predators' search mechanism, more resources distributed around the best member are explored, which means that EPPS could preserve the group diversity without significantly impairing the fast-converging feature and thus both its global searchability and the local searchability are enhanced.

3.4.5 Comparison Between EPPS and Other State-of-the-Art Algorithms

3.4.5.1 Comparison of EPPS with Seven Algorithms on Functions $f_1 - f_{14}$

This section presents a comparative study of EPPS with other seven state-of-the-art algorithms on functions $f_1 - f_{14}$. These algorithms are backtracking search optimization algorithm (BSA) (Civicioglu 2013b), cuckoo search Algorithm (CK) (Civicioglu and Besdok 2013), artificial cooperative search (ACS) (Civicioglu 2013a), strategy adaptation based differential evolution algorithm (SADE) (Qin et al. 2009), differential search algorithm (DSA) (Civicioglu 2012), biogeography based optimization (BBO) (Simon 2008), and comprehensive learning particle swarm optimizer (CLPSO) (Liang et al. 2006). Their experimental results on functions $f_1 - f_{14}$, which were reported in references (Civicioglu 2013a, b), respectively, are directly adopted for comparison in this section. In order to have a fair comparison, the maximum

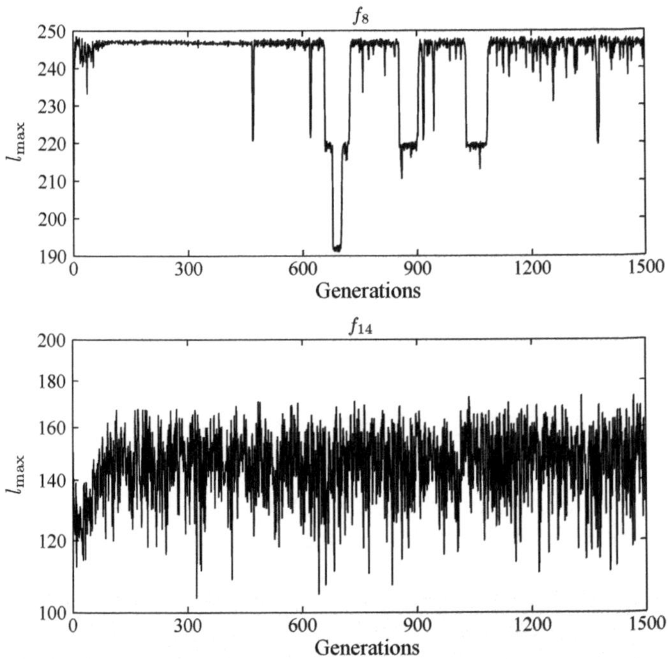

Fig. 3.6 Convergence of l_{max} evaluated on functions f_8 and f_{14}, respectively

number of function evaluations is set to 2,000,000, which is the same as that suggested in Civicioglu and Besdok (2013), Civicioglu (2013a). The comparisons of the mean value between EPPS and other seven algorithms are listed in Table 3.6.

It can be seen from Table 3.6 that EPPS outperforms SADE, CLPSO, BBO, CK, DSA, ACS, and BSA on functions f_4, f_6, f_9, and f_{11}, and has the same performance with these algorithms on functions f_1, f_2, f_5, f_7, f_{13}, and f_{14}. EPPS surpasses CLGSO, BBO, DSA, and ACS on functions f_3 and f_8. As for function f_{12}, SADE, BBO, and BSA cannot find the global optimum. In addition, SADE also cannot find the global optimum on function f_{10}.

3.4.5.2 Comparison of EPPS with Two Algorithms on Functions $f_{15} - f_{20}$

iCMAES-ILS (Liao and Stutzle 2013) and NBIPOP-aCMA (Loshchilov 2013) are within the rank of the top two algorithms for rotated and shifted benchmark functions (Liao and Stutzle 2013). By comparing EPPS with these two algorithms, we can validate the performance of EPPS more comprehensively.

Table 3.7 lists the simulated results obtained by EPPS, NBIPOP-aCMA and iCMAES-ILS, including the best, worst, median, mean, and standard deviation. The

Table 3.6 Comparison of EPPS with BSA, CK, ACS, SADE, DSA, BBO, and CLPSO on benchmark functions $f_1 - f_{14}$. All results have been averaged over 30 runs

f	SADE	CLPSO	BBO	CK	DSA	ACS	BSA	EPPS
1	0	0	0	0	0	0	0	0
2	0	0	0	0	0	0	0	0
3	0	3.2611	7.7318	0	3.8238E-10	1.6643E-11	0	0
4	0.6667	7.1037E-4	0.6673	6.7783E-3	4.3093E-10	0.3333	0.6444	0
5	0	0	0	0	0	0	0	0
6	1.5317E-3	1.1305E-3	5.5095E-4	1.2437E-3	5.8586-3	1.4076E-3	1.9955E-3	0
7	0	0	0	0	0	0	0	0
8	2.1984E-2	2.7530	71.9104	0	0	0	0.3987	0
9	–12569.4866	–12214.1716	–12569.4836	–12569.4866	–12569.4866	–12569.4866	–12569.4866	–12569.4882
10	0.9950	0	0	0	0	0	0	0
11	0.9313	8.0000E-15	9.0000E-16	4.4000E-15	2.2200E-14	8.0000E-15	1.0500E-14	**8.8818E-16**
12	1.7239E-2	0	2.5964E-2	0	0	0	4.9307E-4	0
13	0	0	0	0	0	0	0	0
14	0	0	0	0	0	0	0	0

experimental results obtained by iCMAES-ILS and NBIPOP-aCMA, which were reported in Liao and Stutzle (2013), Loshchilov (2013), respectively, are directly adopted for comparison in this chapter, because the codes of these algorithms are not available. From Table 3.7, it is clearly seen that EPPS performs better than the other two algorithms on functions f_{16}, f_{17}, and f_{20}. EPPS has the same best fitness values as iCMAES-ILS and NBIPOP-aCMA on functions f_{15}, f_{18}. As for function f_{19}, the best fitness value of EPPS is inferior to that of iCMAES-ILS and NBIPOP-aCMA, respectively.

3.4.6 *Application of EPPS to Three Real-World Problems*

3.4.6.1 Solving Unit Commitment Problem

Unit commitment is a significant practical task for power system operation and plays an important role in the deregulated electricity markets. The unit commitment problem in a power system refers to finding a unit commitment schedule that minimizes the commitment and dispatch costs, subject to various constraints (Lee et al. 2014). This problem is commonly formulated as a complex nonlinear and mixed-integer combinational optimization problem with a series of prevailing equality and inequality constraints (Zhao et al. 2013). Moreover, the number of combinations of 0–1 variables grows exponentially as being a large-scale problem. Therefore, the problem is considered as one of the most difficult problems in power system.

Table 3.7 Comparison of EPPS with iCMAES-ILS and NBIPOP-aCMA on benchmark functions $f_{15} - f_{20}$. All results have been averaged over 30 runs

f	Algorithms	Best	Worst	Median	Mean	Std
15	NBIPOP-aCMA	**0**	0	0	0	0
	iCMAES-ILS	**0**	0	0	0	0
	EPPS	**0**	0	0	0	0
16	NBIPOP-aCMA	20.80	21.01	20.95	20.94	4.80×10^{-2}
	iCMAES-ILS	20.80	21.00	20.90	20.90	6.23×10^{-2}
	EPPS	**20.0000**	20.0011	20.0000	20.0006	0.0002
17	NBIPOP-aCMA	0.40	7.63	2.77	3.30	1.83
	iCMAES-ILS	7.10×10^{-2}	8.06	4.53	4.34	1.72
	EPPS	$\mathbf{3.4222 \times 10^{-2}}$	4.1381	0	0.9014	0.7817
18	NBIPOP-aCMA	**0**	0	0	0	0
	iCMAES-ILS	**0**	0	0	0	0
	EPPS	**0**	0	0	0	0
19	NBIPOP-aCMA	$\mathbf{1.40 \times 10^{-2}}$	2.78	4.10×10^{-2}	0.44	0.93
	iCMAES-ILS	1.48×10^{-2}	1.21	0.43	0.38	0.27
	EPPS	5.7074×10^{-2}	0.6915	0.2172	0.2827	0.3108
20	NBIPOP-aCMA	11.12	13.64	13.13	12.94	0.60
	iCMAES-ILS	12.10	15.00	14.50	14.40	0.74
	EPPS	**9.0741**	12.3264	11.2537	11.4117	1.2309

In this section, in order to investigate the capability of EPPS to solve practical problems, it is used to optimize a 10-unit system from literature (Kazarlis et al. 1996). The mathematical formulation of unit commitment is shown as follows (Yang et al. 2015):

$$\begin{cases} \text{minimize} & F = \sum_{t=1}^{T} \sum_{i=1}^{N} [f_i(P_i^t) + ST_i^t(1 - u_i^{t-1})] u_i^t \\ \text{subject to} & \sum_{i=1}^{N} P_i^t u_i^t = P_D^t \\ & \sum_{i=1}^{N} P_{i\max}^t u_i^t \geq P_D^t + P_R^t \\ & P_{i\min} \leq P_i^t \leq P_{i\max} \\ & T_{i,\text{ON}}^t \geq T_{i,\text{up}} \\ & T_{i,\text{OFF}}^t \geq T_{i,\text{down}} \end{cases} \tag{3.24}$$

where

$$f_i(P_i^t) = a_i + b_i P_i^t + c_i (P_i^t)^2 \tag{3.25}$$

Table 3.8 Simulation results of 10-unit system with 10% of spinning reserve

Algorithms	Best cost($)	Mean cost($)	Worst cost($)
EP	564,551	565,352	566,231
PSO	564,212	565,103	565,783
IPSO	563,954	564,162	564,579
HPSO	563,942	NA	NA
SA	565,828	565,988	566,260
QEA-UC	563,938	564,012	564,711
IQEA-UC	563,938	563,938	563,938
BCPSO	563,947	564,285	565,002
C&B	563,938	NA	NA
GSA	563,938	564,008	564,241
GSO	563,938.5123	563,938.9536	563,939.2514
EPPS	**563,937.6863**	**563,937.6872**	**563,937.8490**

NA: not available

and

$$ST_i^t = \begin{cases} S_{hi} & \text{if } T_{i,\text{OFF}}^t \leq T_{i,\text{down}} + Tcold_i \\ S_{ci} & \text{if } T_{i,\text{OFF}}^t > T_{i,\text{down}} + Tcold_i \end{cases} \tag{3.26}$$

where f_i is the fuel cost of the ith unit which is taken as quadratic function; $N = 10$ is number of generators; $T = 24$ is total scheduling period; P_i^t is generation of unit i at time t; $u_i^t \in \{0, 1\}$, is ON/OFF status of unit i at time t (ON = 1 and OFF = 0); ST_i^t is start-up cost of unit i at time t; a_i, b_i, c_i represent the unit cost coefficients; S_{hi} is hot start-cost of unit i; S_{ci} is cold start-up cost of unit i; $Tcold_i$ is cold start time of unit i; $T_{i,\text{down}}$ is minimum down time of unit i; $T_{i,\text{OFF}}^t$ is continuous down time of unit i up to time t.

In the equality and inequality constraints of (3.24), P_D^t denotes the system load demand at time t; P_R^t is spinning reserve at time t; $P_{i\min}$ and $P_{i\max}$ are minimum and maximum generation limit of unit i, respectively; $T_{i,\text{ON}}^t$ is continuously up time of unit i up to time t and $T_{i,\text{up}}$ is the minimum up time of unit i.

The scheduling time horizon T is chosen as one day with 24 intervals of one hour each. The spinning reserve requirement is set to be 10% of total load demand. The input data is described in Kazarlis et al. (1996).

The 10-unit system with 10% spinning reserve is considered in this subsection to further demonstrate the effectiveness of the proposed algorithm. The optimum dispatch of committed generating units, fuel cost, start-up cost, and spinning reserve at all the time horizons are shown in Table 3.9. To validate the computational efficiency of the proposed approach, the simulation results of EPPS are compared with those obtained by EP (Juste et al. 1999), PSO (Zhao et al. 2006), IPSO (Zhao et al. 2006), HPSO (Ting et al. 2006), SA (Simopoulos et al. 2006), QEA-UC (Chung et al. 2011), IQEA-UC (Chung et al. 2011), C&B (Zheng et al. 2015), BCPSO (Chakraborty et al. 2012), GSA (Roy 2013), and GSO. From Table 3.8, it is clearly suggested that EPPS is computationally more efficient than the other methods in terms of solution quality.

Table 3.9 Generation schedule of the 10-unit system with 10% of spinning reserve obtained by EPPS for 24 h

Hour	Unit										Operating	Startup	Reserve
	1	2	3	4	5	6	7	8	9	10	cost($)	cost($)	%
1	455	245	0	0	0	0	0	0	0	0	13683	0	30
2	455	295	0	0	0	0	0	0	0	0	14554	0	21.33
3	455	370	0	0	25	0	0	0	0	0	16809	900	26.12
4	455	455	0	0	40	0	0	0	0	0	18598	0	12.84
5	455	390	0	130	25	0	0	0	0	0	20020	560	20.20
6	455	360	130	130	25	0	0	0	0	0	22387	1100	21.09
7	455	410	130	130	25	0	0	0	0	0	23262	0	15.83
8	455	455	130	130	30	0	0	0	0	0	24150	0	11.00
9	455	455	130	130	85	20	25	0	0	0	27251	860	15.15
10	455	455	130	130	162	33	25	10	0	0	30058	60	10.86
11	455	455	130	130	162	73	25	10	10	0	31916	60	10.83
12	455	455	130	130	162	80	25	43	10	10	33890	60	10.80
13	455	455	130	130	162	33	25	10	0	0	30058	0	10.86
14	455	455	130	130	85	20	25	0	0	0	27251	0	15.15
15	455	455	130	130	30	0	0	0	0	0	24150	0	11.00
16	455	310	130	130	25	0	0	0	0	0	21514	0	26.86
17	455	260	130	130	25	0	0	0	0	0	20642	0	33.20
18	455	360	130	130	25	0	0	0	0	0	22387	0	21.09
19	455	455	130	130	30	0	0	0	0	0	24150	0	11.00
20	455	455	130	130	162	33	25	10	0	0	30058	490	10.86
21	455	455	130	130	85	20	25	0	0	0	27251	0	15.15
22	455	455	0	0	145	20	25	0	0	0	22736	0	12.45
23	455	425	0	0	0	20	0	0	0	0	17645	0	10.00
24	455	345	0	0	0	0	0	0	0	0	15427	0	13.75
Total Cost ($) = 563937											559847	4090	

3.4.6.2 Solving Economic Emission Dispatch Problem

In the past few years, the economic emission dispatch (EED) problem has become an important active research area because it considers the pollutant emissions as well as economic advantages. In general, the unit outputs of the best economic dispatch does not lead to minimum pollution emissions and vice versa. Therefore, it could not solve such problem simply by optimizing a single economic dispatch (ED) problem. In this case, the emission dispatch is added as a second objective to the economic dispatch problem which leads to combined economic emission dispatch (CEED) (Venkatesh et al. 2003; Glotić and Zamuda 2015). In consideration of valve loading effects, the characteristic of CEED is mathematically described as non-smooth and non-convex generation objective function with heavy equality as well as inequality constraints. In general, the formulation of CEED problem is expressed as follows:

$$C_T = F(P) + P_\lambda E(P) \tag{3.27}$$

where the total cost function F(\$/h) can then be expressed as follows (Aragón et al. 2015):

$$F(P) = \sum_{i=1}^{N_G} (a_i + b_i P_i + c_i P_i^2 + \left| e_i \times \sin\left(f_i \times (P_{i,\min} - P_i)\right)\right|) \tag{3.28}$$

and the total emission function E(ton/h) is defined in the following equation (Gent and Lamont 1971):

$$E(P) = \sum_{i=1}^{N_G} 10^{-2} \left(\alpha_i + \beta_i P_i + \gamma_i P_i^2 + \xi_i \exp(\eta_i P_i)\right) \tag{3.29}$$

where P_i is the real power output of unit i, a_i, b_i, and c_i are the cost coefficients of unit i, e_i, and f_i are the coefficients of unit i reflecting valve point effects, N_G is the number of units, $P_{i,\min}$ is the minimum generation limit of ith unit, and α_i, β_i, γ_i, ξ_i, and η_i are the emission coefficients of unit i. P is the vector of real power outputs of units and defined as

$$P = [P_1, P_2, \cdots, P_{N_G}]^{\mathrm{T}} \tag{3.30}$$

The $P_\lambda = (P_{\lambda_1}, \cdots, P_{\lambda_{N_G}})$ is the price penalty factor (\$/ton) which is described as follows (Venkatesh et al. 2003)

$$P_{\lambda_i} = \frac{a_i + b_i P_{i,\max} + c_i P_{i,\max}^2 + \left| e_i \times \sin\left(f_i \times (P_{i,\min} - P_{i,\max})\right)\right|}{10^{-2} \left(\alpha_i + \beta_i P_{i,\max} + \gamma_i P_{i,\max}^2 + \xi_i \exp(\eta_i P_{i,\max})\right)} \tag{3.31}$$

where $P_{i,\max}$ is the maximum output of unit i.

The quality and inequality constraints of CEED are given as follows:

$$\sum_{i=1}^{N_G} P_i = P_D \tag{3.32}$$

$$P_{i,\min} \le P_i \le P_{i,\max} \tag{3.33}$$

where P_D is the total load demand.

Here, a 40-generating units with valve point effects and emission is considered. The input datas for this system come from Venkatesh et al. (2003). The best compromising cost of the test system obtained by EPPS is 191582.0515 \$/h and the best compromising solutions are listed in Table 3.10. The simulation results obtained by EPPS are compared to DE (Basu 2011), MBFA (Hota et al. 2010), DE-HS (Sayah et al. 2014), PSO (Omran and Clerc 2011), GSO (He et al. 2009), and the

Table 3.10 Best compromise solution found with EPPS for 40-unit system

Unit	$P_{i,min}$	$P_{i,max}$	Generation	Unit	$P_{i,min}$	$P_{i,max}$	Generation
1	36	114	114	21	254	550	437.3795
2	36	114	114	22	254	550	437.5076
3	60	120	120	23	254	550	438.0195
4	80	190	178.2003	24	254	550	437.9089
5	47	97	97	25	254	550	437.8020
6	68	140	129.4232	26	254	550	437.7312
7	110	300	300	27	10	150	19.5172
8	135	300	299.5459	28	10	150	19.5123
9	135	300	298.6392	29	10	150	19.5183
10	130	300	130	30	47	97	97
11	94	375	307.5167	31	60	190	175.7608
12	94	375	306.9795	32	60	190	175.8295
13	125	500	433.9502	33	60	190	175.7765
14	125	500	409.2836	34	90	200	200
15	125	500	411.5396	35	90	200	200
16	125	500	411.2092	36	90	200	200
17	220	500	452.0986	37	25	110	104.2640
18	220	500	452.1572	38	25	110	104.2654
19	242	550	437.4438	39	25	110	104.2351
20	242	550	437.4653	40	242	550	437.5198
TP			10500	FC			128729.6391
TE			178569.2016	PPF			0.35198
EC			62852.7876	TC			191582.0515

TP: total power generation (MW); FC: fuel cost ($/h);
TE: total emission (ton/h); PPF: price penalty factor ($/ton);
EC: emission cost ($/h); TC: total generation cost ($/h).

comparison results are given in Table 3.11. It is clearly seen that EPPS could provide better results than other methods in minimum, maximum and mean.

Convergence characteristic for minimum compromise cost of EPPS is shown in Fig. 3.7. We can see from the figure that EPPS could converge to the global optimum at very early iterations. The statistical results on CEED over 30 independent runs by EPPS are depicted in Fig. 3.8. From this figure, it is observed that EPPS consistently produces solutions at or very near to the global optimum, indicating a good convergence characteristic.

Table 3.11 Comparison of compromise cost of different methods on 40-unit system

Methods	Minimum total cost ($/h)	Maximum total cost ($/h)	Mean total cost ($/h)
DE	191594.5053	192251.3565	191695.4643
MBFA	190149.1967	NA	NA
DE-HS	191589.5164	191828.5087	191607.8309
PSO	191598.5325	192305.1021	191699.3392
GSO	191588.5563	191852.2013	191611.3926
EPPS	**191582.0515**	**191726.0181**	**191590.4817**

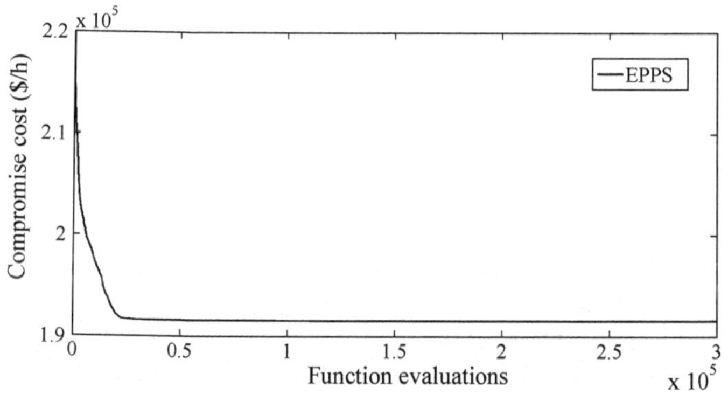

Fig. 3.7 Convergence of EPPS for minimum compromise cost

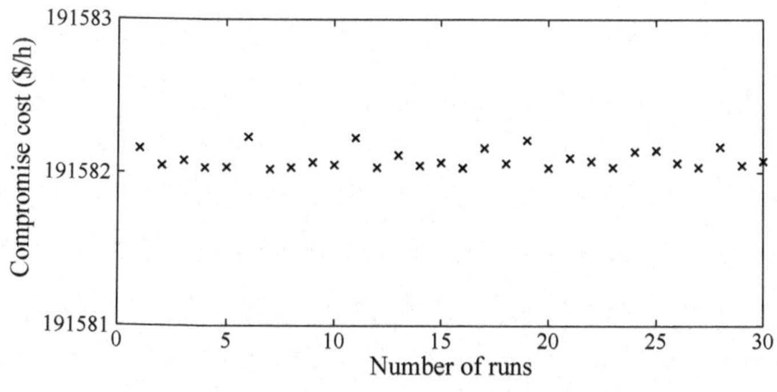

Fig. 3.8 Compromise solution cost obtained by EPPS over 30 trials

3.4.6.3 Estimating the Parameters of an FM Synthesizer

The third real-world problem is to estimate the parameters of an FM synthesizer (Das and Suganthan 2010). It is a highly complex multimodal problem with six parameters, where the vector to be optimized is $\boldsymbol{x} = (a_1, \omega_1, a_2, \omega_2, a_3, \omega_3)$. The fitness function is the summation of square errors between the estimated wave and the target wave as follows:

$$f(\boldsymbol{x}) = \sum_{t=0}^{100} (y(t) - y_0(t))^2 \quad (3.34)$$

where the estimated sound is

$$y(t) = a_1 \cdot \sin(\omega_1 \cdot t \cdot \theta + a_2 \cdot \sin(\omega_2 \cdot t \cdot \theta + a_3 \cdot \sin(\omega_3 \cdot t \cdot \theta))) \quad (3.35)$$

and the target sound is

$$y_0(t) = 1.0 \cdot \sin(5.0 \cdot t \cdot \theta + 1.5 \cdot \sin(4.8 \cdot t \cdot \theta + 2.0 \cdot \sin(4.9 \cdot t \cdot \theta))) \quad (3.36)$$

$\theta = 2\pi/100$ and the parameters are defined in the range $[-6.46.35]$.

The total number of function evaluations is set to 30000 for this problem. Table 3.12 summarizes the minimum, mean, maximum and standard deviation values achieved by EPPS, compared with these of SLPSO (Li et al. 2012), APSO (Zhan et al. 2009), CLPSO (Liang et al. 2006), CPSOH (Li et al. 2012), SPSO (Omran and Clerc 2011), JADE (Li et al. 2012), HRCGA (Li et al. 2012), G-CMA-ES (Li et al. 2012), and GSO. Actually, the minimum value demonstrates the local searchability of the algorithm, the mean value represents the quality of the results obtained by each algorithm, the maximum value reveals the global searchability of the algorithm, and the standard deviation value shows the robustness of the algorithm in optimizing fitness function. It can be seen from Table 3.12 that none of the nine algorithms can find the global optimum for all the 30 independent runs. By observing the minimum values, SLPSO, APSO, SPSO, JADE, HRCGA, and EPPS have found the global optimum at least once in 30 runs, while CLGSO, CPSOH, G-CMA-ES, and GSO did not manage to find the global optimum. As for the maximum value and standard

Table 3.12 Comparison of estimation error of an FM synthesizer

Methods	Minimum value	Maximum value	Mean value	Standard deviation
SLPSO	**0**	**13.79**	4.18	26.99
APSO	**0**	34.22	11.33	41.13
CLPSO	0.007	14.08	3.82	23.53
CPSOH	3.45	42.53	27.08	60.61
SPSO	**0**	18.27	9.88	33.85
JADE	**0**	13.92	7.55	26.18
HRCGA	**0**	17.59	8.41	32.54
G-CMA-ES	3.326	55.09	38.75	**16.77**
GSO	0.002	16.89	8.76	32.21
EPPS	**0**	13.90	**3.69**	23.07

deviation, EPPS ranks second, as SLPSO reaches a smaller value in maximum value and G-CMA-ES reaches a smaller value in standard deviation. Additionally, EPPS performs better than the other nine algorithms in terms of the mean value.

3.5 Summary

This chapter has focused on three multi-objective optimization algorithms, i.e., the MGSO-ACL, MGSOACC, and EPPS. First, the MGSO-ACL consists of three types of group members: producers, scroungers, and rangers. In each generation, the members conferred with the best fitness value of each objective are chosen as the producers, and a number of members are randomly selected as the scroungers, then the rest of members are named the rangers. The MGSO-ACL addresses the adaptive covariance and Lévy flights to increase its exploration and exploitation abilities. Moreover, chaotic search is employed as the rangers' search strategy to maintain the diversity of the group. Chaos is a typical nonlinear phenomenon in nature which is characterized by ergodicity, randomicity, and sensitivity to its initial conditions. Therefore, the MGSOACC is developed utilizing the adaptive covariance and chaotic search. Finally, this chapter has introduced a novel global optimization algorithm, evolutionary predator and prey strategy (EPPS). The EPPS is conceptually simple and easy to implement. To validate its applicability, EPPS has been applied to optimize 20 canonical benchmark functions, including unimodal, multimodal, shifted, and rotated ones, and the results obtained have been compared with those of the other EAs.

References

Aragón V, Esquivel S, Coello CC (2015) An immune algorithm with power redistribution for solving economic dispatch problems. Inf Sci 295:609–632

Auger A, Hansen N (2012) Tutorial CMA-ES: evolution strategies and covariance matrix adaptation. In: GECCO (Companion), pp 827–848

Basu M (2008) Dynamic economic emission dispatch using nondominated sorting genetic algorithm-II. Int J Electric Power Energy Syst 30(2):140–149

Basu M (2011) Economic environmental dispatch using multi-objective differential evolution. Appl Soft Comput 11(2):2845–2853

Chakraborty S, Ito T, Senjyu T, Saber AY (2012) Unit commitment strategy of thermal generators by using advanced fuzzy controlled binary particle swarm optimization algorithm. Int J Electric Power Energy Syst 43(1):1072–1080

Chen WN, Zhang J, Lin Y, Chen N, Zhan ZH, Chung HSH, Li Y, Shi YH (2013) Particle swarm optimization with an aging leader and challengers. IEEE Trans Evol Comput 17(2):241–258

Chung C, Yu H, Wong KP (2011) An advanced quantum-inspired evolutionary algorithm for unit commitment. IEEE Trans Power Syst 26(2):847–854

Civicioglu P (2012) Transforming geocentric cartesian coordinates to geodetic coordinates by using differential search algorithm. Comput Geosci 46:229–247

Civicioglu P (2013a) Artificial cooperative search algorithm for numerical optimization problems. Inf Sci 229:58–76

Civicioglu P (2013b) Backtracking search optimization algorithm for numerical optimization problems. Appl Math Comput 219(15):8121–8144

Civicioglu P, Besdok E (2013) A conceptual comparison of the Cuckoo-search, particle swarm optimization, differential evolution and artificial bee colony algorithms. Artif Intell Rev 39(4):315–346

Conover WJ, Conover W (1980) Practical Nonparametric Statistics. Wiley, New York

Das S, Suganthan P (2010) Problem definitions and evaluation criteria for CEC 2011 competition on testing evolutionary algorithms on real world optimization problems. Jadavpur University, Nanyang Technological University, Kolkata, India

de Athayde Costa e Silva M, Klein CE, Mariani VC, dos Santos Coelho L (2013) Multiobjective scatter search approach with new combination scheme applied to solve environmental/economic dispatch problem. Energy 53(0):14–21

Deb K, Pratap A, Agarwal S, Meyarivan T (2002) A fast and elitist multiobjective genetic algorithm: NSGA-II. IEEE Trans Evol Comput 6(2):182–197

Deb K, Saxena DK (2005) On finding pareto-optimal solutions through dimensionality reduction for certain large-dimensional multi-objective optimization problems. Kangal report 2005011

Derrac J, García S, Molina D, Herrera F (2011) A practical tutorial on the use of nonparametric statistical tests as a methodology for comparing evolutionary and swarm intelligence algorithms. Swarm Evol Comput 1(1):3–18

Dixon AFG (1959) An experimental study of the searching behaviour of the predatory coccinellid beetle Adalia decempunctata. J Animal Ecol 28(2):259–281

Durillo JJ, Nebro AJ, Coello Coello CA, Garcia-Nieto J, Luna F, Alba E (2010) A study of multiobjective metaheuristics when solving parameter scalable problems. IEEE Trans Evol Comput 14(4):618–635

Gent MR, Lamont JW (1971) Minimum emission dispatch. IEEE Trans Power Appar Syst 90(6):2650–2660

Glotić A, Zamuda A (2015) Short-term combined economic and emission hydrothermal optimization by surrogate differential evolution. Appl Energy 141:42–56

Guo CX, Zhan JP, Wu QH (2012) Dynamic economic emission dispatch based on group search optimizer with multiple producers. Electric Power Syst Res 86:8–16

Hansen N, Müller SD, Koumoutsakos P (2003) Reducing the time complexity of the derandomized evolution strategy with covariance matrix adaptation (CMA-ES). Evol Comput 11(1):1–18

Hansen N, Ostermeier A (1996) Adapting arbitrary normal mutation distributions in evolution strategies: the covariance matrix adaptation. In: 1996 Proceedings of IEEE International Conference on Evolutionary Computation. IEEE, pp 312–317

He S, Wu QH, Saunders J (2009) Group search optimizer: an optimization algorithm inspired by animal searching behavior. IEEE Trans Evol Comput 13(5):973–990

Hota P, Barisal A, Chakrabarti R (2010) Economic emission load dispatch through fuzzy based bacterial foraging algorithm. Int J Electric Power Energy Syst 32(7):794–803

Jia DL, Zheng GX, Khan MK (2011) An effective memetic differential evolution algorithm based on chaotic local search. Inf Sci 181:3175–3187

Juste K, Kita H, Tanaka E, Hasegawa J (1999) An evolutionary programming solution to the unit commitment problem. IEEE Trans Power Syst 14(4):1452–1459

Kazarlis SA, Bakirtzis A, Petridis V (1996) A genetic algorithm solution to the unit commitment problem. IEEE Trans Power Syst 11(1):83–92

Lee C, Liu C, Mehrotra S, Shahidehpour M (2014) Modeling transmission line constraints in two-stage robust unit commitment problem. IEEE Trans Power Syst 29(3):1221–1231

Li C, Yang S, Nguyen TT (2012) A self-learning particle swarm optimizer for global optimization problems. IEEE Trans Syst Man Cybern Part B Cybern 42(3):627–646

Liang JJ, Qin AK, Suganthan PN, Baskar S (2006) Comprehensive learning particle swarm optimizer for global optimization of multimodal functions. IEEE Trans Evol Comput 10(3):281–295

Liao TJ, Stutzle T (2013) Benchmark results for a simple hybrid algorithm on the CEC 2013 benchmark set for real-parameter optimization. In: Proceedings of IEEE congress on evolutionary computation, pp 1938–1944

Loshchilov I (2013) CMA-ES with restarts for solving CEC 2013 benchmark problems. In: Proceedings of IEEE congress on evolutionary computation, pp 369–376

Murugan P, Kannan S, Baskar S (2009) Application of NSGA-II algorithm to single-objective transmission constrained generation expansion planning. IEEE Trans Power Syst 24(4):1790–1797

Mustard D (1964) Numerical integration over the n-dimensional spherical shell. Math Comput 18(88):578–589

Niknam T, Narimani MR, Aghaei J, Azizipanah-Abarghooee R (2012) Improved particle swarm optimisation for multi-objective optimal power flow considering the cost, loss, emission and voltage stability index. IET Gener Transm Distrib 6(6):515–527

O'Brien WJ, Evans BI, Howick GL (1986) A new view of the predation cycle of a planktivorous fish, white crappie (pomoxis annularis). Can J Fish Aquat Sci 43(10):1894–1899

Omran MGH, Clerc M (2011) Standard particle swarm optimisation, http://www.particleswarm.info/

Qin AK, Huang VL, Suganthan PN (2009) Differential evolution algorithm with strategy adaptation for global numerical optimization. IEEE Trans Evol Comput 13(2):398–417

Rao PN, Rao KP, Nanda J (1982) An e-coupled fast load flow method. In: Mahalanabis AK (ed) Theory and application of digital control, pp 601–606. Pergamon

Reynolds AM, Smith AD, Reynolds DR, Carreck NL, Osborne JL (2007) Honeybees perform optimal scale-free searching flights when attempting to locate a food source. J Exp Biol 210(21):3763–3770

Roy PK (2013) Solution of unit commitment problem using gravitational search algorithm. Int J Electric Power Energy Syst 53:85–94

Saxena DK, Duro JA, Tiwari A, Deb K, Zhang Q (2013) Objective reduction in many-objective optimization: linear and nonlinear algorithms. IEEE Trans Evol Comput 17(1):77–99

Sayah S, Hamouda A, Bekrar A (2014) Efficient hybrid optimization approach for emission constrained economic dispatch with nonsmooth cost curves. Int J Electric Power Energy Syst 56:127–139

Simon D (2008) Biogeography-based optimization. IEEE Trans Evol Comput 12(6):702–713

Simopoulos DN, Kavatza SD, Vournas CD (2006) Unit commitment by an enhanced simulated annealing algorithm. IEEE Trans Power Syst 21(1):68–76

Storn R, Price K (1997) Differential evolution-a simple and efficient heuristic for global optimization over continuous spaces. J Global Optim 11(4):341–359

Strogatz SH (2014) Nonlinear dynamics and chaos: with applications to physics, biology, chemistry, and engineering. Westview Press

Talatahari S, Azar BF, Sheikholeslami R, Gandomi A (2012) Imperialist competitive algorithm combined with chaos for global optimization. Commun Nonlinear Sci Numer Simul 17:1312–1319

Ting T, Rao M, Loo C (2006) A novel approach for unit commitment problem via an effective hybrid particle swarm optimization. IEEE Trans Power Syst 21(1):411–418

Varadarajan M, Swarup KS (2008) Solving multi-objective optimal power flow using differential evolution. IET Gener Transm Distrib 2(5):720–730

Venkatesh P, Gnanadass R, Padhy NP (2003) Comparison and application of evolutionary programming techniques to combined economic emission dispatch with line flow constraints. IEEE Trans Power Syst 18(2):688–697

Viana EM, de Oliveira EJ, Martins N, Pereira JLR, de Oliveira LW (2013) An optimal power flow function to aid restoration studies of long transmission segments. IEEE Trans Power Syst 28(1):121–129

Viswanathan GM, Buldyrev SV, Havlin S, Luz MGED, Raposo EP, Stanley HE (1999) Optimizing the success of random searches. Nature 401(6756):911–914

Wang LF, Singh CN (2008) Balancing risk and cost in fuzzy economic dispatch including wind power penetration based on particle swarm optimization. Electric Power Syst Res 78(8):1361–1368

Wang H, Yao X (2016) Objective reduction based on nonlinear correlation information entropy. Soft Comput 20(6):2393–2407

Wu QH, Lu Z, Li MS, Ji TY (2008) Optimal placement of facts devices by a group search optimizer with multiple producer. In: 2008 evolutionary computation (IEEE World Congress on Computational Intelligence), CEC 2008. IEEE (2008), pp 1033–1039

Wu QH, Liao HL (2013) Function optimisation by learning automata. Inf Sci 220:379–398

Yang XS (2010) Firefly algorithm, levy flights and global optimization. In: Research and development in intelligent systems XXVI. Springer, London, pp 209–218

Yang L, Jian J, Zhu Y, Dong Z (2015) Tight relaxation method for unit commitment problem using reformulation and lift-and-project. IEEE Trans Power Syst 30:13–23

Zhan ZH, Zhang J, Li Y, Chung HH (2009) Adaptive particle swarm optimization. IEEE Trans Syst Man Cybern Part B Cybern 39(6):1362–1381

Zhao B, Guo CX, Bai BR, Cao YJ (2006) An improved particle swarm optimization algorithm for unit commitment. Int J Electric Power Energy Syst 28(7):482–490

Zhao C, Wang J, Watson JP, Guan Y (2013) Multi-stage robust unit commitment considering wind and demand response uncertainties. IEEE Trans Power Syst 28(3):2708–2717

Zheng JH, Chen JJ, Wu QH, Jing ZX (2015) Multi-objective optimization and decision making for power dispatch of a large-scale integrated energy system with distributed DHCs embedded. Appl Energy 154:369–379

Zheng H, Jian J, Yang L, Quan R (2015) A deterministic method for the unit commitment problem in power systems. Comput Oper Res 1:1–7

Chapter 4
Multi-attribute Decision-Making Support System

Abstract This chapter presents three multi-attribute decision-making support methods, i.e., an improved entropy weight method, evidential reasoning, and interval evidential reasoning. The decision-making methods based on ER and IER approaches are used to determine a final optimal solution from the Pareto-optimal solutions obtained by multi-objective optimization algorithms. The selection of independent evidence for decision-making is investigated together with the study of multiple people and multiple attributes involved in the decision- making process. The decision-making method takes into account both the multiple objectives and the multiple evaluation criteria representing the economy and reliability interests of different operating parties in the LSIES. The performance of these methods are tested in an integrated energy system to find a final operation solution.

Keywords Multi-attribute decision-making support system · Evidential reasoning · Interval evidential reasoning

4.1 Introduction to Decision-Making Support System

The methods for solving MOPs has been classified into three categories by Hwang and Masud (2019) based on the phase in which the DM is involved in the decision-making process: The a priori methods, the interactive methods, and the a posteriori methods. In a priori methods, the DM expresses his/her preference before the optimization process (e.g., setting goals or weights to the objective functions). Nevertheless, the a priori methods are very difficult and abstract for the DM to know beforehand and to be able to accurately quantify his/her preference, either by means of goals or weights. In a posteriori methods, all of the efficient (Pareto optimal) solutions of the problem are generated and then the DM is involved among them to select the most preferred one. This means that the DM has to inspect a large set of solutions to find the most preferred solution, requiring both high computational and cognitive efforts. However, corresponding to MaOPs, the number of all the efficient

Q.-H. Wu et al., *Large-Scale Integrated Energy Systems*, Energy Systems in Electrical Engineering, https://doi.org/10.1007/978-981-13-6943-8_4

solutions is too huge to be selected as the most preferred one. In the interactive methods, phases of dialogue with the DM are interchanged with phases of optimization. The DM progressively drives the search direction with his/her answer toward the most preferred solution. The drawback is that DM never sees the whole picture (the Pareto set) or an approximation of it. Hence, the most preferred solution is "most preferred" in relation to what the DM has seen and compare so far. In fact, the DM prefers to guide the search toward the preferred portion of the Pareto front which is called region of interest (ROI) (Bechikh et al. 2013; Ruiz et al. 2015; Gong et al. 2011) rather than discover the whole Pareto front especially with the increase of the number of objectives.

When the Pareto-optimal set is obtained by the MGSO-ACL, it is necessary for system operators to determine a final optimal solution from the Pareto-optimal solutions. In this chapter, an evidential reasoning (ER) approach (Tang et al. 2004; Chin and Fu 2014) is applied to conduct decision-making for a final optimal solution, with the preference of the operators. Compared with other decision-making methods (Wang and Singh 2008; Niknam et al. 2012; Xiong et al. 2008), the ER makes a decision with adequate evidence fully considering both the multiple objectives and the multiple criteria in order to make full use of the operators' knowledge. Instead of using certain relative weights to multiple objectives as in Guo et al. (2012), the ER takes into account the uncertainties of the operators' cognition. Accordingly, the ER is able to make a convincing decision, determining a final optimal solution which is more preferable for the operators. Furthermore, it has the potential to conduct the ER on the different variable spaces to diminish the region of interest, which can help reduce the computational burden during the decision-making process.

4.2 The IEW Method and Its Decision-Making Model

4.2.1 Relationship Between Two Objectives

Generally speaking, there are three kinds of relationships between objectives, that is, conflict, harmony, and independence (Purshouse and Fleming 2003). When two objectives conflict with each other, it means that the two objectives cannot be improved at the same time. Good values for one implies bad values for the other. The conflicting objectives are considered to be negatively correlated mathematically. On the contrary, a harmony relationship between two objectives means that the improvement of one objective would result in an improvement of the other. Harmonious objectives are denoted in parallel coordinate graphs by noncrossing lines and are considered to be positively correlated mathematically. Therefore, the harmonious objectives can be grouped into a single new compound objective through the simple summation, or we can choose only one of them for optimization. Also, there exists the situation where two objectives are not correlated to each other, that is, they are in independent relationship. As a matter of fact, there are many ways to measure the degree of conflict or harmony between two objectives. In this section, a simple method called Spearman's rank correlation coefficient is utilized to compute correlation degree which is introduced as follows.

4.2.2 *Spearman's Rank Correlation Coefficient*

In statistics, Spearman's rank correlation coefficient or Spearman's rho, named after Charles Spearman and often denoted by the Greek letter ρ, is a nonparametric measure of statistical correlation between two variables (Spearman 1904). It evaluates how well the relationship between two variables can be described using a monotonic function. If there are no repeated data values, a perfect Spearman correlation of $+1$ or -1 occurs when each of the variables is a perfect monotone function of the other.

The Spearman correlation coefficient is defined as Pearson correlation coefficient between ranked variables. Let X_{ij} be the value for objective j in the solution i, then the mathematical formulation of the Spearman's rank correlation coefficient ρ_{ab} between objectives a and b is given below:

$$\begin{aligned} \rho_{ab} &= 1 - \frac{6\sum d_i^2}{n(n^2-1)} \\ d_i &= K_{ia} - K_{ib} \\ K_{ij} &= \text{rank of } X_{ij} \text{ within } X_{.j} \end{aligned} \tag{4.1}$$

where n is the number of solutions and d_i is the difference between ranks. Identical values (rank ties or value duplicates) are assigned a rank equal to the average of their positions in the ascending order of the values. Based on the Spearman correlation coefficient matrix $C^M = \{\rho_{ab}\}$, $(1 \le a \le M, 1 \le b \le M)$, the relation among M objectives can be analyzed.

The sign of the Spearman correlation coefficient ρ_{ab} indicates whether two objectives a and b are in conflict, while the magnitude of ρ_{ab} describes the conflict degree. If ρ_{ab} is a large positive value, the two objectives are highly positively correlated, that is, the relationship between a and b is harmony. If ρ_{ab} is a large negative value, objectives a and b are highly conflicted. If ρ_{ab} is around zero, it indicates that the two objectives are not correlated and are independent with each other. The Spearman correlation coefficient increases in magnitude as a and b become closer to be perfect monotone functions of each other. When they are perfectly monotonically related, $|\rho_{ab}|$ becomes 1.

Although this nonparametric correlation measure involves loss of information as it ignores the specific values of the objectives, it owes advantages over other measures. First, it works without the requirement of comparability between objectives and the objectives can use different units without any conversion. Second, it is useful when we do not have an acknowledgement of the relative importance of each objective, all we want to know is the relationship between them. Third, it is robust and insensitive to any previous normalization. Lastly, the correlation coefficient is easy to compute and can give a clear description of the relationship between two objectives, conflict, harmony, or independence.

4.2.3 *Objective Selection*

With the Spearman correlation coefficient matrix, we can select the most conflicting objectives for MOEAs. Here, we adopt the objective selection method used in Wang and Yao (2015), and the details are shown in Algorithm 1.

Require:
C^M : the Spearman correlation coefficient matrix;
M : the number of objectives;
$S_t = [1 : M]$: a temporary set;
$S_c = \emptyset$: the selected conflicting objectives.
while $S_t \neq \emptyset$ **do**
  if all the elements in C^M are positive **then**
    $J = argmax(sum(C^M(1 : M, j)))$; Find the most representative objective;
  else
    $J = argmin(sum(C^M(i, j)))$, where $C^M(i, j) < 0$ and $1 \leq i \leq M$; Find the most conflicting objective with the remaining objectives.
  end if
  Move f_J from S_t to S_c;
  Add value 0, 1 to $C^M(J, j) > 0$ $(f_j \in S_t)$ as C;
  Classify C into two clusters;
  Delete the objectives in the cluster containing 1 from S_t.
end while
Ensure:
S_c : the selected conflicting objectives.

Algorithm 1: Pseudocode of objective selection

The objective selection algorithm first samples a number of solutions in the feasible space as the learning database and calculates the values of objectives corresponding to the solutions, based on the objective value matrix, the Spearman correlation coefficient matrix can be obtained, then the most conflicting objectives are selected using Algorithm 1.

Entropy is originally the concept in thermodynamics, and it is C. E. Shannon that introduced this concept into the area of information theory, and imposed the entropy a generalized concept (Shannon 2001). Entropy is a measure of uncertainty in the information using probability theory. According to the entropy theory, the quantity and quality of the information for decision-making is one of the important factors to determine the accuracy and reliability of decision-making results. Utilizing the characteristic of entropy that can measure the quantity of the useful information provided by the data, the EW method can be used to assign objective weights. Given the set of alternatives and their attributes, the weight assigned to one attribute by the EW method represents the quantity of information that this attribute conveys to the decision maker, and it also indicates the variance degree among the alternatives with respect to this attribute. A broad distribution represents more uncertainty than a sharply peaked one, so it assigns a small weight to an attribute if it has similar attribute values among alternatives, since such an attribute does not help in differentiating alternatives. The detailed description for EW method is presented as follows:

The decision-making matrix D_{nm} with n alternatives rated on m attributes is first normalized as matrix R_{nm}. According to the definition of entropy, the entropy value of the attribute j is calculated as

$$H_j = -k \sum_{i=1}^{n} f_{ij} \ln f_{ij}, \quad (j = 1, 2, \cdots, m) \tag{4.2}$$

$$f_{ij} = \frac{R_{ij}}{\sum\limits_{i=1}^{n} R_{ij}} \tag{4.3}$$

where $0 \le H_j \le 1$, and it is assumed that if $\ln f_{ij} = 0$, $f_{ij} \ln f_{ij} = 0; i = 1, 2, \cdots, n;$ $j = 1, 2, \cdots, m$.

The entropy weight which is an objective weight for attribute j is thus given by

$$\omega_{ej} = \frac{1 - H_j}{\sum\limits_{j=1}^{m} (1 - H_j)}, \quad (j = 1, 2, \cdots, m) \tag{4.4}$$

where $0 \le \omega_{ej} \le 1$ and $\sum\limits_{j=1}^{m} \omega_{ej} = 1$.

If attribute j owns a subjective weight ω_{sj}, then the comprehensive weight is

$$\omega_j = \frac{\omega_{sj}\omega_{ej}}{\sum_{j=1}^{m} \omega_{sj}\omega_{ej}}, \quad (j = 1, 2, \cdots, m). \tag{4.5}$$

According to (4.4), when all entropy values $H_j \to 1 (j = 1, 2, \cdots, m)$, a delicate difference among the entropy values will bring about the change in the corresponding entropy weight being multiplied. For instance, if the entropy value vectors for attributes are (0.999 9, 0.999 8, 0.999 7) and (0.900 0, 0.800 0, 0.700 0), respectively, the differences among the entropy values of the two vectors are not the same, however, they turn out to have the same entropy weight vector (0.166 7, 0.333 3, 0.500 0) when employing (4.4) for calculation. This manner to assign weights is obviously improper, the reasons are as follows: First, a delicate difference among different attributes means they provide basically the same amount of information, as a result the corresponding entropy weights should be basically the same; Second, different entropy value vectors provide different amounts of information, so they should be given different entropy weights.

To overcome the above disadvantages of the original EW method, we proposed an IEW method, in which the expression of entropy weight is given as

$$\omega_{ej} = \begin{cases} (1-\bar{H})\omega_{ej1} + \bar{H}\omega_{ej2} & H_j < 1 \\ 0 & H_j = 1 \end{cases}, \quad (j = 1, 2, \cdots, m) \tag{4.6}$$

$$\omega_{ej1} = \frac{1 - H_j}{\sum\limits_{j=1, H_j \neq 1}^{m} (1 - H_j)} \qquad \omega_{ej2} = \frac{1/H_j}{\sum\limits_{j=1, H_j \neq 0}^{m} (1/H_j)} \tag{4.7}$$

where $\bar{H}$ is the mean value of the all the entropy values which are not equal to 1, and $0 \leq \omega_{ej} \leq 1$, $\sum_{j=1}^{m} \omega_{ej} = 1$. When all entropy values $H_j \to 0 (j = 1, 2, \cdots, m)$, a delicate difference among the entropy values will result in a multiple change of ω_{ej2}. This means that ω_{ej1} and ω_{ej2} have a complementary relationship. At the same time, the mean value $\bar{H}$ can control the proportions these two weights possessed, that is, when $\bar{H}$ is close to 0, ω_{ej2} will have little contribution to ω_{ej}, and when $\bar{H}$ is close to 1, ω_{ej1} will have little contribution to ω_{ej}. As a result, (4.6) and (4.7) can give a reasonable entropy weight regardless of the extreme entropy values.

Then we apply the IEW method to assess the solutions with multiple objective (attribute) values. The detailed illustration of IEW is given as follows:

(1) Denote the Pareto-optimal solution set obtained by the MOEAs as X_{nm}, it contains n solutions represented by the rows and m objectives represented by the columns. Normalize X_{nm} into R_{nm} as below.

For minimization problem:

$$r_{ij} = \frac{\max\limits_{1 \leq i \leq n} x_{ij} - x_{ij}}{\max\limits_{1 \leq i \leq n} x_{ij} - \min\limits_{1 \leq i \leq n} x_{ij}}, \quad (1 \leq i \leq n, 1 \leq j \leq m) \tag{4.8}$$

For maximization problem:

$$r_{ij} = \frac{x_{ij} - \min\limits_{1 \leq i \leq n} x_{ij}}{\max\limits_{1 \leq i \leq n} x_{ij} - \min\limits_{1 \leq i \leq n} x_{ij}}, \quad (1 \leq i \leq n, 1 \leq j \leq m) \tag{4.9}$$

where r_{ij} denotes the normalized value of the jth objective in the ith solution.

(2) Calculate the entropy value for each objective.

$$H_j = -k \sum_{i=1}^{n} f_{ij} \ln f_{ij}, \quad (j = 1, 2, \cdots, m) \tag{4.10}$$

$$f_{ij} = \frac{R_{ij}}{\sum\limits_{i=1}^{n} R_{ij}} \tag{4.11}$$

(3) Employ the IEW method to calculate the weight ω_j that should be assigned to objective j.

$$\omega_{ej} = \begin{cases} (1-\bar{H})\omega_{ej1} + \bar{H}\omega_{ej2} & H_j < 1 \\ 0 & H_j = 1 \end{cases}, \quad (j = 1, 2, \cdots, m) \tag{4.12}$$

$$\omega_{ej1} = \frac{1 - H_j}{\sum\limits_{j=1, H_j \neq 1}^{m} (1 - H_j)} \qquad \omega_{ej2} = \frac{1/H_j}{\sum\limits_{j=1, H_j \neq 0}^{m} (1/H_j)} \tag{4.13}$$

$$\omega_j = \frac{\omega_{sj}\omega_{ej}}{\sum_{j=1}^{m} \omega_{sj}\omega_{ej}}, \quad (j = 1, 2, \cdots, m) \tag{4.14}$$

where ω_{sj} is the subjective weight the decision maker gives to objective j.

(4) Calculate the aggregation assessment value u_i for each solution using below equation.

$$u_i = \sum_{j=1}^{m} \omega_j r_{ij} \quad (i = 1, 2, \cdots, n) \tag{4.15}$$

(5) Rank the solutions according to the aggregation assessment values, and the larger value, the better solution.

The proposed IEW method can be used for decision-making without the interference of the decision maker, despite that the decision maker can offer the subjective weights for objectives. The IEW method makes full use of the information contained in the solutions, so it is quite simple but rather effective, as there is nothing than the solution itself that can express its quality.

4.3 Evidential Reasoning

The evidential reasoning approach is utilized to determine a final optimal solution for a certain operating point. Consisting of multi-attribute analysis, multi-evidence reasoning, and utility evaluation, the ER combines multiple evidence and provides a suitable mechanism to map the assessment grades of utility evaluation. The multi-attribute analysis, multi-evidence reasoning, and utility evaluation are related to the Pareto-optimal solutions of the power dispatch problem. The calculation of the ER is based on the values of the Pareto-optimal solutions. The overall framework of the ER is shown in Fig. 4.1, and the detailed explanations of the corresponding steps are discussed in detail as follows.

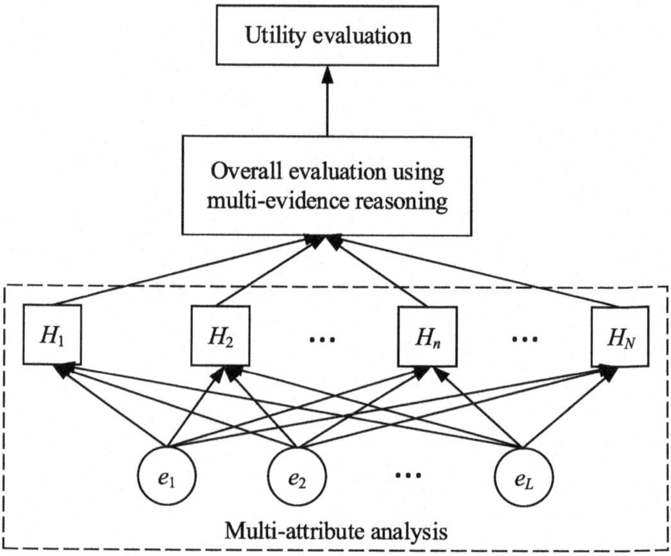

Fig. 4.1 The framework of the evidential reasoning approach

4.3.1 Multi-attribute Analysis

A three-level hierarchy of attributes is taken into consideration in this chapter, with the overall utility evaluation addressed at the top level and a set of basic attributes addressed at the bottom level, which is shown in Fig. 4.1. The set of basic attributes is given as follows:

$$\boldsymbol{E} = \{e_1, e_2, \cdots, e_i, \cdots, e_L\} \tag{4.16}$$

where e_i is the i criterion assigned with a normalized weight $\boldsymbol{\omega} = [\omega_1, \cdots, \omega_i, \cdots, \omega_L](0 < \omega_i < 1)$. The weight ω_i represents the relative importance of criterion e_i, and $\sum_{i=1}^{L} \omega_i = 1$, where L is the total number of criteria. Considering the power dispatch problem formulated in this chapter, the basic attributes can be the objectives of the power grid: fuel cost F_{grid1}, power loss F_{grid3}, voltage stability F_{grid3}, voltage deviation F_{grid4}, or the objective of the distributed DHCs F_{DHCs}.

In order to assess the state of a certain attribute, a set of predefined evaluation grades shown in Fig. 4.1 is presented:

$$\boldsymbol{H} = \{H_1, H_2, \cdots, H_n, \cdots, H_N\} \tag{4.17}$$

where N is the number of evaluation grades. The set of the evaluation grades is set as $\boldsymbol{H} = \{\text{poor, indifference, average, good, excellent, uncertain}\}$, which is assigned with six grades for each attribute.

The basic assessment $\boldsymbol{S}(e_i)$ for attribute e_i is shown as the following distribution of degree of beliefs with regard to different levels of evaluation grades, which is generated based on the preference of the system operators.

$$S(e_i) = \{(H_n, \beta_{n,i}), n = 1, \cdots, N, i = 1, \cdots, L\} \tag{4.18}$$

which means that the attribute e_i is assessed to the grade H_n with a degree of belief $\beta_{n,i}(\beta_{n,i} \geq 0$ and $\sum_{n=1}^{N} \beta_{n,i} \leq 1)$. The assessment $S(e_i)$ is complete if $\sum_{n=1}^{N} \beta_{n,i} = 1$, and it is incomplete if $\sum_{n=1}^{N} \beta_{n,i} < 1$. The case when $\sum_{n=1}^{N} \beta_{n,i} = 0$ denotes a complete lack of information on e_i. For example, the subjective judgement of the system operators for a certain attribute of an alternative is 70% of {good}, 10% of {excellent}, the left 10% of {uncertain} due to the uncertainties of the operators' cognition. Then, we can get the basic assessment for this attribute as the following distribution $S(e) = \{$(poor, 0), (indifference, 0), (average, 0), (good, 0.8), (excellent, 0.1), (uncertain, 0.1)$\}$.

4.3.2 Multi-evidence Reasoning

The ER processes the aggregation based on multiple evidence. A basic probability mass $m_{n,i}$ represents the degree to the ith basic attribute e_i supporting the assessment of the general attribute y with the nth evaluation grade H_n. The mass $m_{n,i}$ is calculated by $m_{n,i} = \omega_i \beta_{n,i}$. The remaining probability mass $m_{H,i}$ is unassigned to any individual grade and decomposed into two parts $\bar{m}_{H,i} = 1 - \omega_i$ and $\tilde{m}_{H,i} = \omega_i(1 - \sum_{n=1}^{N} \beta_{n,i})$.

Let $m_{n,I(i)}$ be the combined probability mass calculated by aggregating the first i (from 1 to i) assessments. $\bar{m}_{n,I(i)}$ and $\tilde{m}_{n,I(i)}$ are the remaining probability masses unassigned to any individual grade after aggregation of the first i assessments. Assuming $m_{n,I(1)} = m_{n,1}, \bar{m}_{n,I(1)} = \bar{m}_{n,1}, \tilde{m}_{n,I(1)} = \tilde{m}_{n,1}$, the combination of the i assessment with $i + 1$ assessment can be depicted by the following recursive expression:

$$\begin{aligned} \{H_n\} : m_{n,I(i+1)} &= K_{I(i+1)}[m_{n,I(i)} m_{n,i+1} \\ &\quad + m_{H,I(i)} m_{n,i+1} + m_{n,I(i)} m_{H,i+1}] \end{aligned} \tag{4.19}$$

$$\begin{aligned} m_{H,I(i)} &= \bar{m}_{H,I(i)} + \tilde{m}_{H,I(i)} \\ \{H\} : \tilde{m}_{H,I(i+1)} &= K_{I(i+1)}[\tilde{m}_{H,I(i)} \tilde{m}_{H,i+1} \\ &\quad + \bar{m}_{H,I(i)} \tilde{m}_{H,i+1} + \tilde{m}_{n,I(i)} \bar{m}_{H,i+1}] \end{aligned} \tag{4.20}$$

$$\bar{m}_{H,I(i+1)} = K_{I(i+1)}[\bar{m}_{H,I(i)} \bar{m}_{H,i+1}] \tag{4.21}$$

where $K_{I(i+1)}$ is defined as below for $i = 1, 2, \cdots, L - 1$

$$K_{I(i+1)} = \left[1 - \sum_{t=1}^{N} \sum_{j=1, j \neq t}^{N} m_{t,I(i)} m_{j,i+1}\right]^{-1} \tag{4.22}$$

After the aggregation of all the basic attributes, the combined degrees of belief β_n corresponding to the grade H_n, and the unassigned degree of belief β_H representing the incompleteness and uncertainty of the overall assessment, are calculated by the following normalization formulation:

$$\{H_n\} : \beta_n = \frac{m_{n,I(L)}}{1 - \bar{m}_{H,I(L)}} \tag{4.23}$$

$$\{H\} : \beta_H = \frac{\tilde{m}_{n,I(L)}}{1 - \bar{m}_{H,I(L)}} \tag{4.24}$$

As a consequence, the overall assessment for the general attribute y can be updated by the distribution of degree of beliefs $\widetilde{\boldsymbol{S}}(y)$ in (4.18) with regard to different evaluation grades:

$$\begin{aligned}\widetilde{\boldsymbol{S}}(y) &= \boldsymbol{S}(e_1) \oplus \boldsymbol{S}(e_2) \cdots \oplus \boldsymbol{S}(e_i) \oplus \cdots \oplus \boldsymbol{S}(e_L) \\ &= \{(H_n, \widetilde{\beta}_n), n = 1, \cdots, N\}\end{aligned} \tag{4.25}$$

where $\oplus$ denotes the aggregation of two attributes. Accordingly, we can get the overall assessment of each alternative by aggregation of the basic assessment for each attribute according to the preference of the operators.

4.3.3 Utility Evaluation

When the distributed descriptions are not sufficient to make the evidential decision, the concept of expected utility has been proposed (Yang and Xu 2002). In this case, the utility of the attribute y can be used for ranking alternatives, which is given as follows:

$$u(y) = \sum_{n=1}^{N} \beta_n u(H_n) \tag{4.26}$$

where it can be determined that as alternative a is preferred to another alternative b if and only if $u(y(a)) > u(y(b))$.

However, in most cases, the assessment for a basic attribute is incomplete, which means $\beta_H > 0$. Furthermore, three measures are addressed to characterize the assessment for attribute y, i.e., the minimum, maximum, and averaged utilities which are formulated as

$$u_{\max}(y) = \sum_{n=1}^{N-1} \beta_n u(H_n) + (\beta_N + \beta_H) u(H_N) \tag{4.27}$$

$$u_{\min}(y) = (\beta_1 + \beta_H)u(H_1) + \sum_{n=2}^{N} \beta_n u(H_n) \tag{4.28}$$

$$u_{\text{avg}}(y) = \frac{(u_{\max}(y) + u_{\min}(y))}{2} \tag{4.29}$$

The alternative having the lower utility evaluation ranks lower, while the alternative having the higher utility evaluation ranks higher. As a consequence, the ranking between two alternatives a and b can be assigned based on their minimum, maximum, and averaged utility evaluation. If $u_{\min}(y(a)) > u_{\max}(y(b))$, then a is preferred to b; if and only if $u_{\min}(y(a)) = u_{\min}(y(b))$ and $u_{\max}(y(a)) = u_{\max}(y(b))$, then a is indifferent to b; otherwise, averaged utility is applied to generate a ranking between a and b.

4.4 Interval Evidential Reasoning

The framework of the IER is shown in Fig. 6.4. As shown in this figure, the IER consists of three hierarchical steps: multi-attribute analysis, multi-evidence reasoning and utility evaluation. The detailed explanations of the corresponding steps are discussed in detail as follows (Fig. 4.2).

4.4.1 Multi-attribute Analysis

Suppose a MADA problem has M alternatives $a_m, m = 1, \cdots, M$, and a set of L basic attributes addressed at the bottom level for each alternative. The set of basic attributes is given as

$$\boldsymbol{E} = \{e_1, e_2, \cdots, e_l, \cdots, e_L\} \tag{4.30}$$

where $\boldsymbol{E}$ is a set of attributes which is assigned with a normalized weight $\boldsymbol{\omega} = [\omega_1, \cdots, \omega_l, \cdots, \omega_L](0 \le \omega_l \le 1)$. The weight ω_l represents the relative importance of attribute e_l, and $\sum_{l=1}^{L} \omega_l = 1$.

Different from the original ER algorithm, the performance of each alternative can be assessed to an individual grade or a grade interval. Accordingly, the complete set of all individual grades and grade intervals for assessing each attribute, denoted by $\widehat{H}$, can be represented by

$$\widehat{\boldsymbol{H}} = \{H_{pq} | \; p = 1, \cdots, N, q = p, \cdots, N\} \tag{4.31}$$

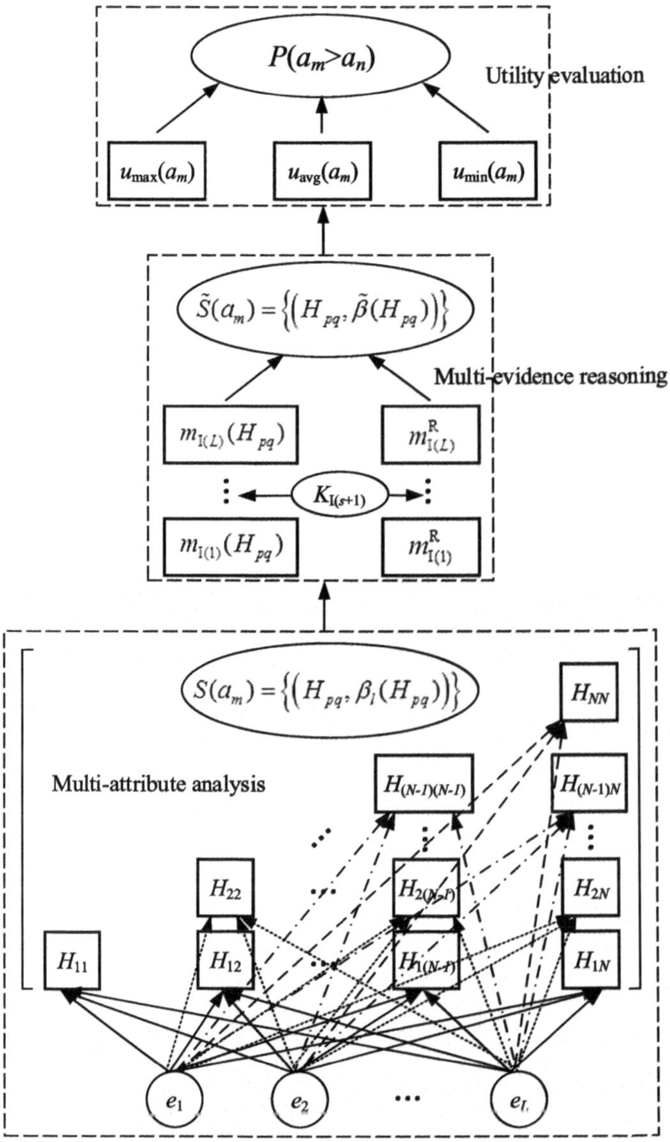

Fig. 4.2 The framework of the interval evidential reasoning

or equivalently

$$\widehat{\boldsymbol{H}} = \begin{bmatrix} H_{11} & H_{12} \cdots & H_{1(N-1)} & H_{1N} \\ & H_{22} \cdots & H_{2(N-1)} & H_{2N} \\ & \ddots & \vdots & \vdots \\ & & H_{(N-1)(N-1)} & H_{(N-1)N} \\ & & & H_{NN} \end{bmatrix} \tag{4.32}$$

where H_{pp}, $(p = 1, \cdots, N)$ denote an individual grade while H_{pq}, $p = 1, \cdots, N$; $q = p + 1, \cdots, N$ denotes the interval grade that is the union of the individual grades H_{pp} and $H_{(p+1)(p+1)}, \cdots, H_{qq}$.

Suppose alternative a_m is assessed on the basic attributes e_l using the assessment grades H_{pq}. The basic assessment can be represented as follows:

$$S(a_m) = \{(H_{pq}, \beta_l(H_{pq}))|\ l = 1, \cdots, L, H_{pq} \in \widehat{\boldsymbol{H}}\} \tag{4.33}$$

where $\beta_l(H_{pq})$ is the basic belief degree based on the subjective judgment of the system operators, which means that the attribute e_l is assessed to the grade H_{pq} with a degree of belief $\beta_l(H_{pq})(\beta_l(H_{pq}) \geq 0$ and $\sum_{H_{pq} \in \widehat{\boldsymbol{H}}} \beta_l(C) = 1)$.

4.4.2 *Multi-evidential Reasoning*

Similarly to the original ER algorithm, the IER also processes the aggregation based on the multiple evidence. A basic probability mass function $m_l(H_{pq})$ assigned to each element in $\{\widehat{\boldsymbol{H}}\}$ is defined as

$$m_l(H_{pq}) = \omega_l \beta_l(H_{pq}), \quad l = 1, \cdots, L, H_{pq} \in \widehat{\boldsymbol{H}} \tag{4.34}$$

Besides, the remaining mass function m_l^{R}, which is to be assigned depending on the relative importance of other attributes, can be calculated by

$$\begin{aligned} m_l^{\mathrm{R}} &= 1 - \sum_{H_{pq} \in \widehat{H}} m_l(H_{pq}) \\ &= 1 - \omega_l \Big(\sum_{H_{pq} \in \widehat{H}} \beta_l(H_{pq}) \Big) = 1 - \omega_l \end{aligned} \tag{4.35}$$

where m_l^{R} represents the remaining role that other attributes can play in the assessment, which should eventually be assigned to the assessment grades in a way dependent on the importance of other attributes.

According to the D-S theory and the original ER aggregation rules, the recursive aggregation of the IER can be illustrated as follows.

Let $s, s = 1, 2, \cdots, L$ be the recursive number and the combination results of the first s^{th} attributes be denoted as $m_{\mathrm{I}(s)}(\cdot)$. In the first recursion $s = 1$, we have

$$m_{\mathrm{I}(1)}(H_{pq}) = m_1(H_{pq}),\ H_{pq} \in \widehat{\boldsymbol{H}} \tag{4.36}$$

$$m_{\mathrm{I}(1)}^{\mathrm{R}} = m_1^{\mathrm{R}} \tag{4.37}$$

In the next recursion $s + 1$, we have

$$\begin{aligned} m_{\mathrm{I}(s+1)}(H_{pq}) = K_{\mathrm{I}(s+1)}\Big\{ & - m_{\mathrm{I}(s)}(H_{pq})m_{s+1}(H_{pq}) \\ & + \sum_{i=1}^{p}\sum_{j=q}^{N}[m_{\mathrm{I}(s)}(H_{ij})m_{s+1}(H_{pq}) \\ & + m_{\mathrm{I}(s)}(H_{pq})m_{s+1}(H_{ij})] \\ & + \sum_{i=1}^{p-1}\sum_{j=q+1}^{N}[m_{\mathrm{I}(s)}(H_{iq})m_{s+1}(H_{pj}) \\ & + m_{\mathrm{I}(s)}(H_{pj})m_{s+1}(H_{iq})] \\ & + m_{\mathrm{I}(s)}^{\mathrm{R}}m_{s+1}(H_{pq}) \\ & + m_{\mathrm{I}(s)}(H_{pq})m_{s+1}^{\mathrm{R}}\Big\} \end{aligned} \tag{4.38}$$

$$m_{\mathrm{I}(s+1)}^{\mathrm{R}} = K_{\mathrm{I}(s+1)}m_{\mathrm{I}(s)}^{\mathrm{R}}m_{s+1}^{\mathrm{R}} \tag{4.39}$$

where $K_{\mathrm{I}(s+1)}$ is the scaling factor used to make sure that $\sum_{H_{pq}\in\widehat{\boldsymbol{H}}} m_{\mathrm{I}(s+1)}(H_{pq}) + m_{\mathrm{I}(s+1)}^{\mathrm{R}} = 1$, which is calculated by

$$\begin{aligned} K_{\mathrm{I}(s+1)} = \Big\{1 - \sum_{p=1}^{N}\sum_{q=p}^{N}\sum_{i=1}^{p-1}\sum_{j=i}^{p-1}&[m_{\mathrm{I}(s)}(H_{ij})m_{s+1}(H_{pq}) \\ + m_{\mathrm{I}(s)}(H_{pq})m_{s+1}(H_{ij})]\Big\}^{-1} \end{aligned} \tag{4.40}$$

Note that in (4.38) and (4.40), the summing up calculation $\sum_{i=i_1}^{i_2} f(i)$ will not be carried out when $i_1 > i_2$. Mathematically, we may say that $\sum_{i=i_1}^{i_2} f(i) = 0$ when $i_1 > i_2$, and this convention applies throughout this section.

Applying the previous aggregation process recursively until all the L attribute assessments are aggregated, the overall assessment of the alternative a_m can be expressed as follows:

$$\begin{aligned} \widetilde{S}(a_m) &= S(e_1 \oplus e_2 \cdots \oplus e_l \oplus \cdots \oplus e_L) \\ &= \{(H_{pq}, \widetilde{\beta}(H_{pq})),\ H_{pq} \in \widehat{\boldsymbol{H}}\} \\ \text{with }\ \widetilde{\beta}(H_{pq}) &= \frac{m_{\mathrm{I}(L)}(H_{pq})}{1 - m_{\mathrm{I}(L)}^{\mathrm{R}}} \end{aligned} \tag{4.41}$$

where $\oplus$ denotes the aggregation of two attributes, and $\widetilde{\beta}(H_{pq})$ is the overall assessment assigned to the corresponding grade H_{pq} after the aggregation based on the multiple evidence.

4.4.3 Utility Evaluation

When the distributed descriptions are not sufficient to make the evidential decision, the concept of expected utility has been proposed (Yang and Xu 2002). In this case, the expected utility evaluations of the alternative a_m can be used for ranking alternatives. Suppose $u(H_{pp})$ is the value of the assessment grade H_{pp} with $u(H_{p+1,p+1}) > u(H_{pp})$ as it is assumed that the grade $H_{p+1,p+1}$ is preferred to H_{pp}. Due to the interval uncertainty, the maximum, minimum, and average expected utilities are calculated for ranking alternatives. As the belief degree $\widetilde{\beta}(H_{pq})$ can be assigned to the best grade in the interval H_{pq} which is H_{qq}, if the uncertainty turns out to be favorable to the assessed alternative, the the maximum utility evaluation can be expressed as

$$u_{\max}(a_m) = \sum_{p=1}^{N}\sum_{q=p}^{N}\widetilde{\beta}(H_{pq})u(H_{qq}) \tag{4.42}$$

Similarly, in the worst case, if the interval uncertainty turns out to be against the assessed alternative, i.e., the belief degree $\widetilde{\beta}(H_{pq})$ assigned to H_{pp}, the worst grade in the interval H_{pq}, then the minimum expected utility evaluation can be calculated as

$$u_{\min}(a_m) = \sum_{p=1}^{N}\sum_{q=p}^{N}\widetilde{\beta}(H_{pq})u(H_{pp}) \tag{4.43}$$

Accordingly, the average of the expected utility evaluation is given by

$$u_{\text{avg}}(a_m) = \frac{(u_{\max}(a_m) + u_{\min}(a_m))}{2} \tag{4.44}$$

The alternative having the lower utility evaluation ranks lower, while the alternative having the higher utility evaluation ranks higher. As a consequence, the ranking between two alternatives a_m and a_n can be assigned based on their minimum, maximum, and averaged utility evaluations. If $u_{\min}(a_m) > u_{\max}(a_n)$, then a_m is preferable to a_n to an extent of 100%; if and only if $u_{\min}(a_m) = u_{\min}(a_n)$ and $u_{\max}(a_m) = u_{\max}(a_n)$, then a_m is indifferent to a_n; otherwise, the averaged utility evaluation is applied to generate a ranking between a_m and a_n. In this case, the extent to which the interval number $[u_{\min}(a_m), u_{\max}(a_m)] \geq [u_{\min}(a_n), u_{\max}(a_n)]$ is calculated using the following equation:

$$P(m > n) = \frac{\max(0, u_{m2} - u_{n1}) - \max(0, u_{m1} - u_{n2})}{(u_{m2} - u_{m1}) + (u_{n2} - u_{n1})} \tag{4.45}$$

where u_{m1} and u_{m2} are the minimum and maximum utility evaluations of alternative m, while u_{n1} and u_{n2} are the minimum and maximum utility evaluations of alternative n, respectively.

4.5 Simulation Studies

Simulation studies are carried out on the modified IEEE 30-bus system with distributed DHCs and wind power generation integrated. The distributed energy units are interconnected via the 30-bus power grid, hence the test system can be treated as a LSIES. The detailed locations of generators, electricity loads, wind turbines, and distributed DHCs are summarized in Table 4.1.

As for the wind turbines, the rated, cut-in and cut-out wind speeds are set as 12.5 m/s, 4.0 m/s and 20 m/s, respectively. In addition, the detailed parameter settings of the DHC system can be found in our previous work (Jiang et al. 2014; Jing et al. 2014).

Applying the MGSO-ACL, the multi-objective optimization problem with one objective addressed from the power grid and the other addressed from the distributed DHCs, is optimized with comparisons with the original GSOMP. Figure 4.3 shows the best Pareto fronts obtained by MGSO-ACL and GSOMP. As shown in the figure, it is obvious that MGSO-ACL outperforms GSOMP in terms of searching for better converged and more evenly distributed non-dominated solutions.

4.5.1 The ER Method Performance Test

As for the Pareto-optimal solutions obtained by MGSO-ACL, the decision-making method, evidence reasoning, is applied to determine a final optimal solution in order to balance the interests of the power grid and the distributed DHCs. Taking the Pareto-optimal solutions of the competing objectives $[F_{\text{grid1}}, F_{\text{DHCs}}]$ as an example, thus the objectives $[F_{\text{grid2}}, F_{\text{grid3}}, F_{\text{grid4}}]$ are treated as the multiple evaluation criteria for decision-making. In order to comprehensively present the ER, five Pareto-optimal

Table 4.1 The detailed locations of different components in the modified IEEE 30-bus system

Components	Node	Components	Node
Generators	1, 2, 5, 13	Wind farms	8, 11
Distributed DHCs	7, 12, 21, 30	Electricity loads	the others

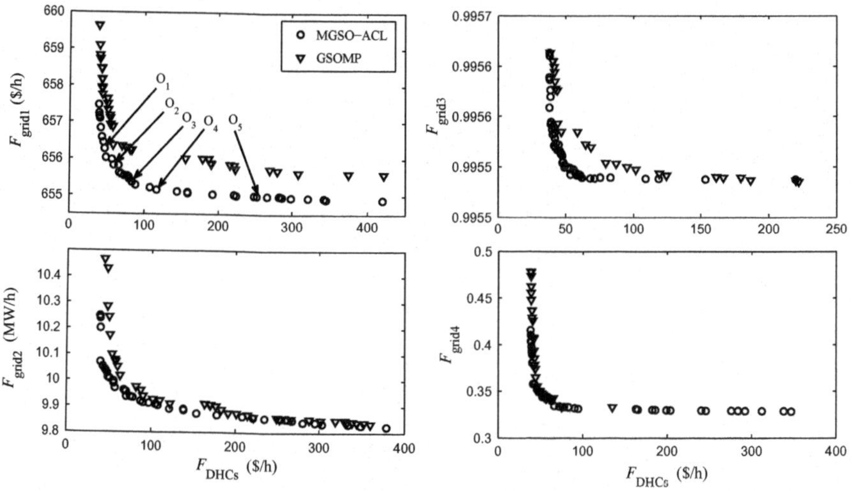

Fig. 4.3 Pareto fronts obtained by MGSO-ACL and GSOMP for different objectives of power grid and DHCs

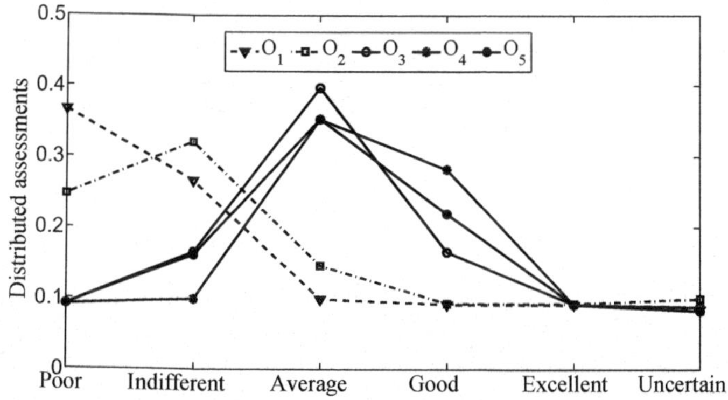

Fig. 4.4 The distributed assessments for the five selected solutions

solutions obtained by MGSO-ACL are selected, namely [O_1, O_2, O_3, O_4, O_5], which are shown in Fig. 4.3.

The corresponding values of the objectives and the evaluation criteria of the five selected Pareto-optimal solutions are shown in Table 4.2. According to the results shown in the table, it is difficult to simply choose a final optimal solution since the objectives and evaluation criteria are in conflict with each other and none of the five Pareto-optimal solutions are superior in terms of all the objectives and evaluation criteria. Hence, the ER which takes into account all the multiple objectives and the multiple evaluation criteria is applied to tackle the complex decision-making problem.

Table 4.2 The values of the objectives and the evaluation criteria of the five selected Pareto-optimal solutions

	Objectives	Evaluation criteria
Solution	$[F_{grid1}, F_{DHCs}]$	$[F_{grid2}, F_{grid3}, F_{grid4}]$
O_1	[656.43, **44.85**]	[15.97, 0.996, 1.991]
O_2	[655.82, 58.76]	[14.79, 0.995, 1.956]
O_3	[655.36, 83.38]	[15.25, **0.990**, 1.989]
O_4	[655.21, 107.46]	[14.71, 0.991, **1.737**]
O_5	[**654.99**, 248.76]	[**14.56**, 0.997, 1.761]

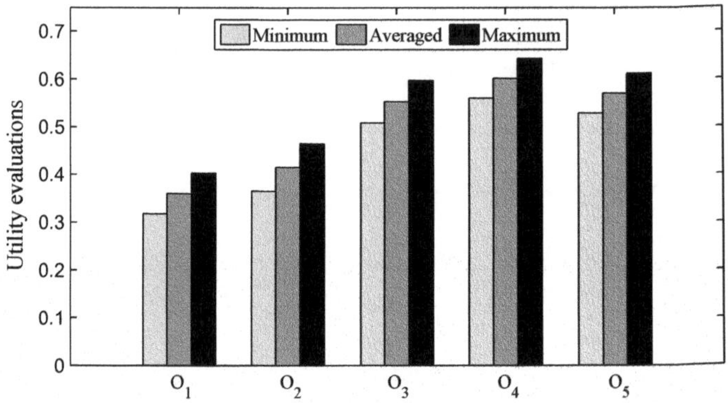

Fig. 4.5 Utility evaluations for the five selected solutions

Suppose the set of relative weights ω_i of the evaluation criteria $[F_{grid2}, F_{grid3}, F_{grid4}]$ is $\omega = [0.2, 0.4, 0.4]$. Because the objectives $[F_{grid1}, F_{DHCs}]$ represent the economy benefits for the power grid and distributed DHCs, the evaluation criteria representing the interests of reliability operation should be assigned with greater weights. In addition, the set of evaluation grades is set as $\boldsymbol{H}$ = {poor, indifference, average, good, excellent, uncertain}. Therefore, the subjective judgement can be set based on the ranking of the evaluation criteria of the five selected Pareto-optimal solutions. Note that without loss of generality, the evaluation grade $\boldsymbol{H}$(uncertain) of the subjective judgement is set as 0.1 for all the selected Pareto-optimal solutions. Applying the multi-evidence reasoning of the ER, the assessments for the five selected Pareto-optimal solutions by aggregating power loss, voltage stability, and voltage deviations are, therefore, presented by the distributed assessments, which are shown in Fig. 4.4 and Table 4.3.

According to the distributed assessments for the Pareto-optimal solutions, it is still not straightforward to distinguish them from each other, because the degrees of their beliefs assessed to the evaluation grades are of no great significance. Hence, the utility evaluation is applied in order to rank the solutions.

Table 4.3 The distributed assessments for the five selected solutions

Solution	Poor	Indifferent	Average	Good	Excellent	Uncertain
O_1	**0.3660**	0.2661	0.0990	0.0918	0.0918	0.0852
O_2	0.2469	**0.3208**	0.1460	0.0932	**0.0932**	**0.0998**
O_3	0.0921	0.1653	**0.3970**	0.1653	0.0921	0.0884
O_4	0.0917	0.0990	0.3527	**0.2820**	0.0917	0.0828
O_5	0.0919	0.1610	0.3532	0.2191	0.0919	0.0829

Therefore, the utility evaluations for the five Pareto-optimal solutions $[O_1, O_2, O_3, O_4, O_5]$ are presented in Fig. 4.5. As shown in the figure, it is evident that the ranking of the five selected solutions is $O_4 \succ O_5 \succ O_3 \succ O_2 \succ O_1$. Hence, as for the objectives $[F_{grid1}, F_{DHCs}]$ and the evaluation criteria $[F_{grid2}, F_{grid3}, F_{grid4}]$, O_4 can be chosen as the final optimal solution which is determined with the preference of system operators.

Considering other decision-making methods, like the fuzzy decision method (Niknam et al. 2012; Xiong et al. 2008), the best compromised solution obtained by these methods only considers the values of the multiple objectives of the Pareto-optimal solutions. In this way, O_3 is selected as the final solution. However, the other evaluation criteria of O_3 in terms of F_{grid2} and F_{grid4} perform worst for power grid among the five alternatives. Thus, it can be concluded that the final solution obtained by the fuzzy decision method considering only the multiple objectives is not convincing enough. Furthermore, according to the Pareto-dominance principle in (3.4), it is worthwhile to mention that each Pareto-optimal solution is an optional choice, so multiple evaluation criteria apart from the objectives should be taken into consideration. As a consequence, compared with the fuzzy decision method, ER is more pragmatic for solving the complex decision-making problem of the optimal power dispatch representing the competing interests of the power grid and the DHCs interconnected in the LSIES.

In terms of the objectives $[F_{grid2}, F_{DHCs}]$, $[F_{grid3}, F_{DHCs}]$ and $[F_{grid4}, F_{DHCs}]$, applying the same decision-making procedure of the ER aforementioned, a final optimal solution can also be selected from the corresponding Pareto-optimal solutions.

4.5.2 *The IEW Method Performance Test*

To verify the rationality of the IEW method in assigning proper weights, we adopt different types of entropy values, and compare the results with other methods, as shown in Table 4.4. Here for convenience, we denote the original method as A, the method in Zhou et al. (2007) as B, the method in Lu et al. (2015) as C, and our method as D. It can be seen that when the entropy value vectors are too large or too small,

Table 4.4 The entropy weights given by four different entropy weight methods against different types of entropy values

Value type	Entropy value	A	B	C	D
$H_j \to 1$	0.999 9	0.166 7	0.333 3	0.333 2	0.333 3
	0.999 8	0.333 3	0.333 3	0.333 3	0.333 3
	0.999 7	0.500 0	0.333 4	0.333 5	0.333 4
	0.900 0	0.166 7	0.296 3	0.166 7	0.267 9
	0.800 0	0.333 3	0.333 3	0.333 3	0.330 5
	0.700 0	0.500 0	0.370 4	0.500 0	0.401 6
$H_j \to 0.5$	0.600 0	0.266 7	0.289 9	0.266 7	0.268 5
	0.500 0	0.333 3	0.333 3	0.333 3	0.328 8
	0.400 0	0.400 0	0.377 8	0.400 0	0.402 7
	1.000 0	0.000 0	0.156 6	0.000 0	0.000 0
	0.500 0	0.357 1	0.333 3	0.357 1	0.300 0
	0.100 0	0.642 9	0.511 1	0.642 9	0.700 0
$H_j \to 0$	0.300 0	0.291 7	0.277 8	0.291 7	0.269 7
	0.200 0	0.333 3	0.333 3	0.333 3	0.321 2
	0.100 0	0.375 0	0.388 9	0.375 0	0.409 1
	0.000 3	0.333 3	0.333 3	0.333 3	0.333 3
	0.000 2	0.333 3	0.333 3	0.333 3	0.333 3
	0.000 1	0.333 4	0.333 4	0.333 4	0.333 4

our method can give more moderate and reasonable entropy weight vectors. When the entropy values are all close to 0.5, the difference between the weight vectors given by our IEW method and the original EW method is tiny, both of them are reasonable. In particular, compared with other methods, our method gives a weight vector which has more discrimination among the attributes for entropy value vector (1.000 0, 0.500 0, 0.100 0).

Take the Pareto-optimal solution set consisting of 39 solutions and 5 objectives obtained by the multi-objective optimization in Zheng et al. (2015) as the decision-making data. Table 4.5 gives the Pareto-optimal solution set, the rank of solutions given by the IEW method (rank1), and the rank by the PROMETHEE (rank2). It can be seen that these two methods give basically the same ranking schemes and both choose the 19th solution as the best solution, indicating the effectiveness of the IEW method in decision-making. The decision-making method used in Zheng et al. (2015) is ER, however, this method is complex and need the decision maker to evaluate the solutions elaborately. So it only chooses the five solutions in boldface from Table 4.5 for decision-making and only considers the last three objectives in the decision-making process. We employ the IEW method and the PROMETHEE method to rank these five solutions, and compare the results with the rank scheme given by the ER (rank3). As shown in Table 4.6, all of these three methods give the same rank scheme. However, the IEW method explicitly makes use of the quality

Table 4.5 The rank schemes given by IEW and PROMETHEE for the solution set consisting of 39 solutions and 5 objectives

Solution	Objective1	Objective2	Objective3	Objective4	Objective5	Rank1	Rank2
1	654.89	420.09	14.51	0.996	1.745	28	28
2	654.91	345.51	14.58	0.996	1.739	22	22
3	654.93	342.57	14.53	0.996	1.756	23	23
4	654.94	321.18	14.54	0.995	1.757	12	13
5	654.95	298.69	14.52	0.996	1.743	16	16
6	654.97	286.94	14.54	0.997	1.760	25	24
7	654.98	285.11	14.53	0.996	1.76	17	17
8	654.99	283.12	14.53	0.996	1.765	18	18
9	654.99	267.22	14.56	0.996	1.755	15	15
10	655.00	252.83	14.56	0.996	1.759	14	14
11	**655.00**	**248.77**	**14.56**	**0.997**	**1.761**	**21**	**20**
12	655.00	224.01	14.55	0.996	1.755	9	9
13	655.03	220.90	14.54	0.996	1.757	10	10
14	655.03	192.11	15.56	0.996	1.760	30	30
15	655.07	158.82	14.57	0.996	1.758	6	5
16	655.10	158.48	14.57	0.996	1.77	8	7
17	655.11	144.50	14.58	0.995	1.773	2	2
18	655.15	116.61	14.63	0.996	1.777	7	6
19	**655.21**	**107.46**	**14.71**	**0.991**	**1.737**	**1**	**1**
20	655.29	88.00	14.62	0.996	1.786	5	4
21	**655.36**	**83.38**	**15.25**	**0.990**	**1.989**	**4**	**8**
22	655.45	79.55	14.70	0.996	1.812	11	11
23	655.51	78.68	14.72	0.996	1.858	19	19
24	655.53	76.77	14.72	0.996	1.819	13	12
25	655.54	73.04	15.76	0.996	1.838	32	32
26	655.58	68.94	14.69	0.994	1.839	3	3
27	655.62	66.17	14.78	0.996	1.894	24	25
28	655.81	65.39	14.71	0.996	1.920	27	27
29	**655.82**	**58.76**	**14.79**	**0.995**	**1.956**	**26**	**26**
30	655.97	56.38	14.85	0.996	1.916	29	29
31	656.02	48.35	14.94	0.996	1.958	31	31
32	656.26	46.52	14.86	0.993	1.951	20	21
33	**656.44**	**44.85**	**15.97**	**0.996**	**1.971**	**37**	**37**
34	656.57	42.75	15.04	0.996	1.945	33	33
35	656.83	41.10	15.13	0.996	1.996	34	34
36	657.09	40.02	15.17	0.996	2.120	38	38
37	657.15	39.25	15.04	0.996	2.035	35	35
38	657.23	39.01	15.11	0.996	2.058	36	36
39	657.46	38.49	15.25	0.996	2.174	39	39

Table 4.6 The rank schemes given by IEW, PROMETHEE, and ER for the solution set consisting of 5 solutions and 3 objectives

Solution	Objective3	Objective4	Objective5	Rank1	Rank2	Rank3
11	14.71	0.991	1.737	5	5	5
19	14.71	0.991	1.737	4	4	4
21	15.25	0.990	1.989	3	3	3
29	14.79	0.995	1.956	1	1	1
33	15.97	0.996	1.971	2	2	2

information of the solution represented by its numerical value, and doesn't need the decision maker to allocate the solutions into several evaluation degrees according to their numerical value and then transform the evaluation degrees into utility value like the ER. That is why the IEW method is suitable for online running. It can pick up the final solution for decision maker automatically just using the information given by the Pareto-optimal solution set.

4.6 Summary

This chapter has concentrated on three multi-attribute decision analysis support methods, i.e., the IEW, ER, and IER which consists of multi-attribute analysis, multi-evidence reasoning, and utility evaluation. The decision-making methods can combine the complex multiple evaluation criteria to map the assessment grades of utility evaluation. Compared with other decision-making methods, the presented multi-attribute decision analysis support methods can make full utilization of the multiple objectives including those which are not included for optimization directly. Moreover, the multi-attribute decision analysis support methods can also take into account the uncertainty of the operators' cognition as a quantifiable evaluation grade. It is demonstrated the multi-attribute decision analysis support methods are effective in making a convincing decision to satisfy the economy and reliability interests of the power grid and the distributed DHCs.

References

Bechikh S, Said LB, Ghdira K (2013) Group preference-based evolutionary multi-objective optimization with non-equally important decision makers: application to the portfolio selection problem. Int J Comput Inf Syst Ind Manag Appl 5(1)

Chin KS, Fu C (2014) Integrated evidential reasoning approach in the presence of cardinal and ordinal preferences and its applications in software selection. Expert Syst Appl 41(15):6718–6727

Gong M, Liu F, Zhang W, Jiao L, Zhang Q (2011) Interactive MOEA/D for multi-objective decision making. In: Genetic and evolutionary computation conference, GECCO 2011, Proceedings, Dublin, Ireland, July, pp 721–728

Guo CX, Zhan JP, Wu QH (2012) Dynamic economic emission dispatch based on group search optimizer with multiple producers. Electric Power Syst Res 86:8–16

Hwang CL, Masud ASM (2019) Multiple objective decision making methods and applications: a state-of-the-art survey, vol. 164. Springer Science & Business Media

Jiang XS, Jing ZX, Li YZ, Wu QH, Tang WH (2014) Modeling and operation optimization of an integrated energy based direct district water-heating system. Energy 64:375–388

Jing ZX, Jiang XS, Wu QH, Tang WH, Hua B (2014) Modelling and optimal operation of a small-scale integrated energy based district heating and cooling system. Energy 73:399–415

Lu YL, Chen JS, Qi J, Ji P, Zhou JZ (2015) Multi-objective decision making method for reservoir flood operation based on the improved entropy weight and set pair analysis. Water Res Power 33(1):43–46

Niknam T, Narimani MR, Aghaei J, Azizipanah-Abarghooee R (2012) Improved particle swarm optimisation for multi-objective optimal power flow considering the cost, loss, emission and voltage stability index. IET Gener Transm Distrib 6(6):515–527

Purshouse RC, Fleming PJ (2003) Conflict, harmony, and independence: relationships in evolutionary multi-criterion optimisation. In: Evolutionary multi-criterion optimization. Springer, Heidelberg, pp 16–30

Ruiz AB, Saborido R, Luque M (2015) A preference-based evolutionary algorithm for multiobjective optimization: the weighting achievement scalarizing function genetic algorithm. J Glob Optim 62(1):101–129

Shannon CE (2001) A mathematical theory of communication. ACM SIGMOBILE Mobile Comput Commun Rev 5(1):3–55

Spearman C (1904) The proof and measurement of association between two things. Am J Psychol 15(1):72–101

Tang WH, Spurgeon K, Wu QH, Richardson ZJ (2004) An evidential reasoning approach to transformer condition assessments. IEEE Trans Power Deliv 19(4):1696–1703

Wang LF, Singh CN (2008) Balancing risk and cost in fuzzy economic dispatch including wind power penetration based on particle swarm optimization. Electric Power Syst Res 78(8):1361–1368

Wang H, Yao X (2015) Objective reduction based on nonlinear correlation information entropy. Soft Comput 1–15

Xiong HG, Cheng HZ, Li HY (2008) Optimal reactive power flow incorporating static voltage stability based on multi-objective adaptive immune algorithm. Energy Conv Manag 49(5):1175–1181

Yang JB, Xu DL (2002) On the evidential reasoning algorithm for multiple attribute decision analysis under uncertainty. IEEE Trans Syst Man Cybern Part A Syst Hum 32(3):289–304

Zheng JH, Chen JJ, Wu QH, Jing ZX (2015) Multi-objective optimization and decision making for power dispatch of a large-scale integrated energy system with distributed DHCs embedded. Appl Energy 154:369–379

Zhou HC, Zhang GH, Wang GL (2007) Multi-objective decision making approach based on entropy weights for reservoir flood control operation. J Hydraul Eng 38(1):100–106

Chapter 5
Planning of the Large-Scale Integrated Energy Systems

Abstract This chapter presents the planning problems of the LSIES considering the optimal unit sizing and the multi-stage contingency-constrained co-planning, respectively. First, a comprehensive framework including a multi-objective interval optimization model and evidential reasoning (ER) approach is introduced to solve the unit sizing problem of small-scale integrated energy systems, with uncertain wind and solar energies integrated. In the multi-objective interval optimization model, interval variables are introduced to tackle the uncertainties of the optimization problem. Aiming at simultaneously considering the cost and risk of a business investment, the average and deviation of life cycle cost (LCC) of the integrated energy system are formulated. Second, a multi-stage contingency-constrained co-planning for electricity-gas systems (EGS) interconnected with gas-fired units and power-to-gas (P2G) plants considering the uncertainties of load demand and wind power. The MCC model considers the long-term co-planning for EGS with the short-term operation constraints, while enabling systems to satisfy N-1 reliability criterion. These planning problems are solved utilizing the multi-objective optimization algorithms and decision-making support methods introduced in the previous chapters.

Keywords Planning problems · Optimal unit sizing · Multi-stage contingency-constrained

5.1 Introduction to Planning of the LSIES

Integrated energy systems (IESs) provide a valuable opportunity to efficiently and sustainably utilize the energy, due to its ability to accommodate different energy systems, involving electricity, gas, heating, and transport (Mancarella 2014; Wang et al. 2017). Specially, the synergies between the electricity and gas systems are dramatically increasing in recent years, in order to improve the efficiency and compliance with the stringent environmental regulations (Zheng et al. 2017). According to the monthly energy review provided by the U.S. Energy Information Administration (EIA) U.S. Energy Information Administration (2019), the U.S. natural gas consumption by electric sector has increased from 32% in 2007 to 39% in 2017.

Q.-H. Wu et al., *Large-Scale Integrated Energy Systems*, Energy Systems in Electrical Engineering, https://doi.org/10.1007/978-981-13-6943-8_5

Furthermore, the new gas-fired generating units are expected to be projected for the next decade, due to their low emissions and high efficiency. Consequently, the necessity for devoting significant efforts to the coordinated expansion planning for electricity systems and natural gas infrastructure continues to grow.

Increasing research interest has been directed towards the co-optimization expansion planning of integrated electricity-gas systems, involving generation expansion planning (GEP), transmission expansion planning (TEP) and natural gas grid expansion planning (NGGEP). Since TEP and GEP can seriously influence NGGEP, these three items are always simultaneously accomplished. In this context, a chance-constrained and reliability programming optimization model for combined GEP and NGGEP is reported in Odetayo et al. (2018). In Hemmati et al. (2017), coordination of GEP, TEP and energy storage systems in microgrids is analyzed by using a strong meta-heuristic optimization algorithm. A two-stage robust optimization problem of TEP and NGGEP is formulated in Shao et al. (2017) for enhancing the power grid resilience. Hu et al. (2016) carries out a NSGA-II multi-objective optimization for TEP and NGGEP. In Barati et al. (2015), a multi-period integrated GEP+TEP+NGGEP framework for large-scale integrated energy systems is established. Qiu et al. (2016) proposes a linear programming approach for expansion co-planning in gas and electricity markets to minimize the total capital and operational cost, while satisfying the reliability and security requirements. Qiu et al. (2015) develops a novel model on multi-stage flexible expansion co-planning under uncertainties in a combined electricity and gas market. However, in most of the previous research, only gas-fired generators are served as the linkage between electricity and gas systems. Power-to-gas (P2G) plants (Colbertaldo et al. 2018) are not taken into consideration. As an indispensable medium to transform the surplus wind power into hydrogen or synthetic natural gas (SNG) (Parra and Patel 2016), P2G technology has a direct influence on the collaborative planning and operation dispatch of electricity-gas systems.

Although Guandalini et al. (2015) carries out the capacity configuration optimization of P2G plants and gas turbines for improving wind energy dispatchability, the security constraints of gas network are not considered. The collaborative site-selection planning problem is investigated in Moskalenko et al. (2014), but the detailed mathematical models are not given. The authors of Zeng et al. (2017) focus on the coordinated expansion planning of P2G plants and gas-fired power generation and propose a hybrid algorithm to solve the bilevel programming problem. However, the capacity expansion of transmission lines and gas pipelines is not included. Wang et al. (2015) proposes a multi-stage co-planning of electricity-gas systems and shows that P2G plants expansion planning (P2GEP) will influence the planning solution of TEP. Nevertheless, neither did it consider the uncertainties of load demand and renewable wind energy, nor consider N-1 contingency.

Based on the above premises, this section presents a multi-stage contingency-constrained co-planning (MCC) model for electricity-gas systems (EGS) interconnected with gas-fired units and power-to-gas plants considering the uncertainties of load demand and wind power. The proposed MCC model considers the long-term co-planning for electricity-gas systems with the short-term operation constraints, including operational constraints of power system, natural gas fuel supply availability, physical limits of P2G plants and transmission N-1 contingency constraints. The objective is to minimize the total co-planning cost (including investment cost, operation cost and curtailed wind energy cost) across the planning horizon.

The proposed MCC is formulated as a mixed-integer linear programming (MILP) problem, which can be solved by outer linearization algorithm (OLA) (Duran and Grossmann 1986), branch and bound (BB) (Kelley 1960), and Benders decomposition (BD) (Benders 1962). Due to the large size of the relaxed master program, the work per iteration in the OLA is often great, which augments the computational burden. The authors of Majidi-Qadikolai and Baldick (2016) adopt BB algorithm to solve the TEP problem under just two different operation states. However, with the increase of operation states, the number of decision variables raises in multiples and the MILP problem becomes very hard to be solved. BD algorithm is applied in Zhang et al. (2015), Roh et al. (2007) to decompose the initial co-optimization planning model into several subproblems. Nevertheless, to evaluate all the possible contingencies of transmission lines has tremendously increased the number of iterations.

In this section, an iterative Benders decomposition (IBD) method is presented to solve the proposed MCC model. At the first step, in order to accommodate the intermittent characteristic of load demand and wind power energy, we adopt affinity propagation (AP) (Frey and Dueck 2007) clustering to determine a set of representative scenarios. AP is a clustering algorithm based on the similarities between data points and proves to be fast and efficient for large-scale data sets. Compared with other cluster methods, like k-means, the main advantage of AP is its ability to determine the number of clustered scenarios without pre-specification. Afterward, the IBD method divides the initial MCC model into one master investment problem and three subproblems (including pre-contingency check subproblem, N-1 check subproblem and operation optimal subproblem). Furthermore, to reduce the number of iterations and improve the computational efficiency, a contingency screening method based on line outage distribution factors (LODFs) (Guler et al. 2007; Guo et al. 2009) is adopted to quickly detect and create the core contingency set to satisfy the N-1 criterion, due to its straightforward way to calculate post-contingency power flows. The iterative process will continue between master problem and each subproblem until an economic, reliable, and fuel supply feasible solution is obtained, by adding the corresponding Benders cut constraints. The final co-planning solution offers electricity-gas systems planners the optimal capacity, location, installation year of new transmission lines, gas pipelines, gas-fired units, and P2G plants.

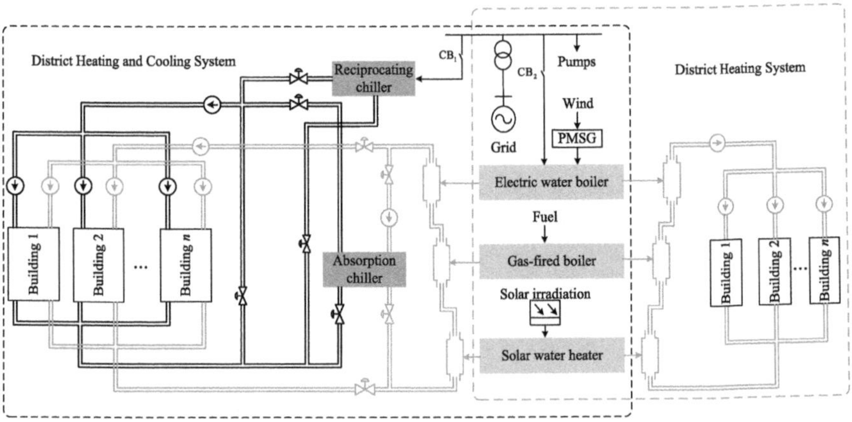

Fig. 5.1 Schematic diagram of two separate systems: a district heating system and a district heating and cooling system

5.2 Optimal Unit Sizing for Integrated Energy Systems

5.2.1 Unit Sizing Problem Formulation

5.2.1.1 System Description

In this section, two small-scale integrated energy systems proposed in our previous works, i.e., a small-scale direct district heating system (DH) (Jiang et al. 2014) and a small-scale district heating and cooling system (DHC) (Jing et al. 2014), are employed to conduct the unit sizing study. Figure 5.1 shows the simplified schematic diagram of the two systems, in which the right-hand side is the DH system and the left-hand side is the DHC system. It should be noted that they are two separate systems working independently.

The DH system integrates wind and solar energies to supply hot water to the residential consumer for space heating (Jiang et al. 2014). The energy supply units consist of a standalone wind turbine generator (WT), a flat-plate solar water heater with solar collectors (SC), an electric water boiler (EB), and a low temperature central gas-fired boiler (GB). The energy captured by WT is transferred to EB for hot water heating. The energies come from the wind energy system and solar water heater are given the first priority to be utilized to heat the water. If the heat generated from the renewable energies cannot meet the heating load, electricity will be imported from the utility grid to EB in order to cover the shortage. Meanwhile, GB will consume natural gas to generate heat as well. Noted that the electricity consumed by various water pumps is imported from the grid.

The DHC system makes joint use of wind energy, solar energy, and conventional fossil fuel energies to supply the residential consumer with space heating in the heating season and to provide space cooling for them in the cooling season (Jing et al. 2014). In the heating season, the DHC system operates in the heating mode, the operation strategy of which is the same as that of the DH system. In the cooling season, the absorption chiller (AC) utilizes the thermal energy generated by EB (driven by WT), GB, and the solar water heater while the reciprocating chiller (RC) is driven by the electricity directly imported from the utility grid. Considering the higher transfer efficiency of RC than that of AC, the circuit breaker CB_1 is switched on and the CB_2 is switched off in the cooling season. For the detailed descriptions of the two systems, Jiang et al. (2014) and Jing et al. (2014) can be referred.

5.2.1.2 Optimization Problem Formulation

In order to apply the optimization technologies, the mathematical model of the optimization problem should be developed first, which will be described in the following subsections.

a. Decision variables

For the unit sizing problem of the integrated energy system, the decision variables include the rated capacity of each energy supply unit and energy consumption of each conventional fossil fuel unit (Ren et al. 2012), which are categorized as independent and dependent variables in this section, respectively. Aiming at obtaining the optimal unit sizing solution, the independent variables consist of the rated capacity (area) of various energy supply units are expressed as follows.

$$X = [P_{\mathrm{WT}}^{\mathrm{rated}},\ P_{\mathrm{SC}}^{\mathrm{rated}},\ P_{\mathrm{GB}}^{\mathrm{rated}},\ P_{\mathrm{EB}}^{\mathrm{rated}},\ P_{\mathrm{RC}}^{\mathrm{rated}},\ P_{\mathrm{AC}}^{\mathrm{rated}}]^{\mathrm{T}} \tag{5.1}$$

where X denotes a vector of independent variables. $P_{\mathrm{WT}}^{\mathrm{rated}}$, $P_{\mathrm{GB}}^{\mathrm{rated}}$, $P_{\mathrm{EB}}^{\mathrm{rated}}$, $P_{\mathrm{RC}}^{\mathrm{rated}}$ and $P_{\mathrm{AC}}^{\mathrm{rated}}$ represent the rated capacity of WT, GB, EB, RC, and AC, respectively. $P_{\mathrm{SC}}^{\mathrm{rated}}$ represents the total area of SC.

The dependent variables including the energy consumption of conventional energy supply units are presented as follows.

$$Y = [P_{\mathrm{elec}}^{h},\ P_{\mathrm{gas}}^{h}]^{\mathrm{T}} \tag{5.2}$$

where Y denotes a vector of dependent variables. P_{elec}^{h} and P_{gas}^{h} denote the electricity and natural gas consumed by EB/RC and GB during the hth operation hour, respectively.

b. Objective function

Life cycle cost (LCC) is a cumulative cost throughout the lifespan of a project (Abbes et al. 2014), which is often used to indicate the economic feasibility of the project. Therefore, LCC of the integrated energy system will be analyzed in this section. The components of LCC considered in this model can be divided into three parts including the initial investment cost, the operation cost as well as the residual cost. The mathematical model of LCC is formulated as follows:

$$C_{\mathrm{LCC}} = C_{\mathrm{in}} + C_{\mathrm{op}} + C_{\mathrm{rc}} \tag{5.3}$$

where C_{LCC} denotes the life cycle cost of the integrated energy system. C_{in} denotes the initial investment cost. C_{op} denotes the total operation cost. C_{rc} represents the residual cost of the energy supply unit.

A. Initial investment cost

The initial investment cost, C_{in}, is the initial amount of funds that should be totally invested in purchasing the energy supply units, which includes the capacity (area) dependent and independent costs, the installation cost and the subsidy of the renewable energy system.

$$C_{\mathrm{in}} = \sum_{i=1}^{I} \left((1 + a_i^{\mathrm{st}}) c_i^{\mathrm{cap}} P_i^{\mathrm{rated}} + f_i^{\mathrm{cap}} + s_i \right) \tag{5.4}$$

where I denotes the total number of energy supply units. P_i^{rated} denotes the rated capacity (area) of the ith energy supply unit. c_i^{cap} denotes the capacity- (area) dependent unit capital cost of the ith equipment. f_i^{cap} represents the capacity (area) independent cost. s_i denotes the subsidy of the renewable energy system. a_i^{st} denotes the installation cost coefficient of the ith energy supply unit which is took as a fraction of its initial capital cost.

B. Operation cost

The operation cost, C_{op}, which should be paid to maintain the daily operation of the system, is formulated as follows:

$$C_{\mathrm{op}} = \sum_{n=1}^{N} \left(C_{\mathrm{mt}}^{n} + C_{\mathrm{elec}}^{n} + C_{\mathrm{gas}}^{n} + C_{\mathrm{em}}^{n} \right) W(d,\ n) \tag{5.5}$$

where N denotes the lifespan of the energy supply units. C_{mt}^{n}, C_{elec}^{n}, C_{gas}^{n} and C_{em}^{n} denote the maintenance cost, the electricity cost, the natural gas cost and the emission cost, respectively. $W(d,\ n)$ denotes the present worth factor of the nth year, which refers to the discount rate d, and it is given by

$$W(d, n) = (1 + d)^{-n} \tag{5.6}$$

The maintenance cost is composed of the fixed and variable ones, which are depended on the capacity of the energy supply unit and the amount of energy produced, respectively (Guo et al. 2013). In addition, as the labor cost takes a large portion of the cost that should be paid during the operation periods, it is also taken into account in this model, which is treated as a constant annual salary according to the economic indicators reported in Dong and Xu (2015). The expression of the maintenance cost is given as follows:

$$C_{\mathrm{mt}}^{n} = \sum_{h=1}^{H} \sum_{i=1}^{I} \left(f_i^{\mathrm{mt}} c_i^{\mathrm{cap}} P_i^{\mathrm{rated}} + a_i^{\mathrm{mt}} P_i^h + a^{\mathrm{salary}} \right) \tag{5.7}$$

where H denotes the total yearly operation hours. f_i^{mt} and a_i^{mt} denote the fixed and variable maintenance cost coefficients, respectively. a^{salary} denotes the labor's salary. P_i^h denotes the hourly energy output of the ith unit.

The electricity cost can be calculated by the hourly electricity tariff multiplied by the amount of electricity consumed by the electricity-driven unit and water pumps during each hour.

$$C_{\mathrm{elec}}^{n} = \sum_{h=1}^{H} c_{\mathrm{elec}}^{h} (P_{\mathrm{elec}}^{h} + P_{\mathrm{pump}}^{h}) \tag{5.8}$$

where c_{elec}^{h} denotes the hourly electricity tariff. P_{pump}^{h} denotes the hourly electricity consumed by various water pumps. The detailed calculation of P_{pump}^{h} can be reffered to Jiang et al. (2014).

The natural gas cost is the cost of the total natural gas comsumed by GB during the operation periods, which can be calculated by the following equation:

$$C_{\mathrm{gas}}^{n} = \sum_{h=1}^{H} c_{\mathrm{gas}} P_{\mathrm{gas}}^{h} \tag{5.9}$$

where c_{gas} denotes the natural gas price.

As the great reduction of pollutant emissions is one of the most remarkable advantages that motivates the development of integrated energy systems, the emission cost is taken into consideration. It is the penalty on the environmental quality reduction and ecological destruction caused by the emissions including CO, CO_2, SO_2, and NO_x from utility electricity and natural gas consumption, which can be calcuated as follows (Ren et al. 2012; Ramakumar et al. 1993):

$$C_{\mathrm{em}}^{n} = \sum_{h=1}^{H} \sum_{g=1}^{G} \gamma^{g} \left(E_{\mathrm{elec}}^{g} (P_{\mathrm{elec}}^{h} + P_{\mathrm{pump}}^{h}) + E_{\mathrm{gas}}^{g} P_{\mathrm{gas}}^{h} \right) \tag{5.10}$$

where G denotes the types of the pollutant emissions. γ^{g} represents the environment value of the gth pollutant emitted. E_{elec}^{g} and E_{gas}^{g} represent the per unit amount of the gth pollutants emitted by electricity and natural gas consumption, respectively.

C. Residual cost

The residual cost, C_{rc}, consists of the decommission cost and the salvage value of the equipment, both of which are the costs that should be paid (or received) at the end of the lifespan. Both of them are considered as fractions of the initial expenditure of each energy supply unit.

$$C_{\mathrm{rc}} = \sum_{i=1}^{I} \left(C_i^{\mathrm{dc}} - C_i^{\mathrm{sv}}\right) W(d, N) \tag{5.11}$$

$$C_i^{\mathrm{dc}} = a_i^{\mathrm{dc}} c_i^{\mathrm{cap}} P_i^{\mathrm{rated}} \tag{5.12}$$

$$C_i^{\mathrm{sv}} = a_i^{\mathrm{sv}} c_i^{\mathrm{cap}} P_i^{\mathrm{rated}} \tag{5.13}$$

where C_i^{dc} and C_i^{dc} denote the decommission cost and salvage value of the ith energy supply unit, respectively. a_i^{dc} and a_i^{sv} denote the fractions of decommission cost and salvage value to the inital expenditure of the ith energy supply unit, respectively.

c. Constraints

In the unit sizing problem of the integrated energy system, the optimization constraints consist of the energy balance constraints and various energy supply units performance and sizing constraints. As has been discussed above, the DHC system operates the same way as the DH system does in the heating mode. Therefore, in order to conduct a clear description, the constraints are divided into two categories, including those of the heating mode and those of the cooling mode.

A. Heating mode

(1). *Constraints on unit sizing*: As for the DH system, the lower and upper bounds of capacity of each unit should be limited in order to avoid over sizing which leads to a profligacy of resources:

$$P_i^{\min} \leq P_i^{\mathrm{rated}} \leq P_i^{\max} \tag{5.14}$$

where $P_i^{\min}$, $P_i^{\max}$ and P_i^{rated} denote the minimum, maximum and rated capacity (area) of the ith energy supply unit, respectively.

In addition, the power generated from the wind energy system is input to EB. Therefore, the rated capacity of WT should not exceed that of EB, which is given by

$$P_{\mathrm{WT}}^{\mathrm{rated}} \leq P_{\mathrm{EB}}^{\mathrm{rated}} \tag{5.15}$$

where $P_{\mathrm{WT}}^{\mathrm{rated}}$ and $P_{\mathrm{EB}}^{\mathrm{rated}}$ denote the rated capacity of WT and EB, respectively.

(2). *Constraints on energy balance*: At the operation stage, the output of the system should balance the heating load at any time interval:

$$\eta_{\mathrm{EB}}(P_{\mathrm{WT}}^h + P_{\mathrm{elec}}^h) + P_{\mathrm{SC}}^h + \eta_{\mathrm{GB}} P_{\mathrm{gas}}^h = Q_{\mathrm{HL}}^h \tag{5.16}$$

where P^h_{WT} and P^h_{SC} denote the hourly output of WT and SC, respectively. η_{EB} and η_{GB} represent the efficiency of EB and GB, respectively. Q^h_{HL} denotes the hourly heating load.

(3). *Constraints on energy supply unit performance*: The ouput of each energy supply unit should not exceed its operation limit.

$$P^{\min}_{\mathrm{GB}} \leq \eta_{\mathrm{GB}} P^h_{\mathrm{gas}} \leq P^{\mathrm{rated}}_{\mathrm{GB}} \tag{5.17}$$

$$P^{\min}_{\mathrm{WT}} \leq P^h_{\mathrm{WT}} \leq P^{\mathrm{rated}}_{\mathrm{WT}} \tag{5.18}$$

$$P^{\min}_{\mathrm{SC}} \leq P^h_{\mathrm{SC}} \leq P^{\mathrm{rated}}_{\mathrm{SC}} \tag{5.19}$$

Meanwhile, the total amount of electricity purchased from the utility grid and imported from the wind energy system should not surpass the rated capacity of EB.

$$P^{\min}_{\mathrm{EB}} \leq \eta_{\mathrm{EB}} (P^h_{\mathrm{WT}} + P^h_{\mathrm{elec}}) \leq P^{\mathrm{rated}}_{\mathrm{EB}} \tag{5.20}$$

B. Cooling mode

(1). *Constraints on unit sizing*: In the cooling mode, besides the capacity (area) constraints on the WT, SC, GB, and EB as shown in (5.14), the rated capacity of RC and AC should also be limited.

(2). *Constraints on operation*: In the cooling season, the cooling load is supplied by the electricity-driven unit RC and the heat-driven unit AC. Therefore, the energy balance constraint can be expressed by

$$\mathrm{COP}_{\mathrm{RC}} P^h_{\mathrm{elec}} + \mathrm{COP}_{\mathrm{AC}} P^h_{\mathrm{AC}} = Q^h_{\mathrm{CL}} \tag{5.21}$$

where $\mathrm{COP}_{\mathrm{RC}}$ and $\mathrm{COP}_{\mathrm{AC}}$ denote the coefficient of performance (COP) of RC and AC, respectively. P^h_{AC} denotes the hourly input energy of AC. Q^h_{CL} represents the hourly cooling load.

In this mode, the circuit breaker CB_2 is switched off, which means that EB is merely supported by WT. Therefore, the energy input into AC is primarily generated by WT, SC, and GB. Hence, (5.21) can be reformulated as follows:

$$\mathrm{COP}_{\mathrm{RC}} P^h_{\mathrm{elec}} + \mathrm{COP}_{\mathrm{AC}} \left(\eta_{\mathrm{EB}} P^h_{\mathrm{WT}} + P^h_{\mathrm{SC}} + \eta_{\mathrm{GB}} P^h_{\mathrm{gas}} \right) = Q^h_{\mathrm{CL}} \tag{5.22}$$

(3). *Constraints on energy supply unit performance*: In addition to the performance constraints on WT, SC, and GB as shown in (5.14)–(5.15), the output of the RC and AC should also be limited with their corresponding operation limit:

$$P^{\min}_{\mathrm{RC}} \leq \mathrm{COP}_{\mathrm{RC}} P^h_{\mathrm{elec}} \leq P^{\mathrm{rated}}_{\mathrm{RC}} \tag{5.23}$$

$$P^{\min}_{\mathrm{AC}} \leq \mathrm{COP}_{\mathrm{AC}} P^h_{\mathrm{AC}} \leq P^{\mathrm{rated}}_{\mathrm{AC}} \tag{5.24}$$

Moreover, as the utility grid is no longer connected to EB in the cooling mode, the constraint (5.20) becomes

$$P_{\mathrm{EB}}^{\min} \le \eta_{\mathrm{EB}} P_{\mathrm{WT}}^{h} \le P_{\mathrm{EB}}^{\mathrm{rated}} \tag{5.25}$$

d. Renewable generations

(1). *Wind energy*: An off-grid wind turbine with PMSG (permanent magnet synchronous generator) is applied in both the DH system and the DHC sytem. The output power of the wind turbine is limited by its rated capacity, as well as its designed parameters, including cut-in, cut-out, and rated wind speed. The mathematical model based on cubic law clearly indicates the relationship between these parameters, which is shown as follows (Thapar et al. 2011):

$$P_{\mathrm{WT}}^{h} = \begin{cases} 0 & 0 \le v_h < v_{\mathrm{ci}} \\ \frac{v_{\mathrm{ci}}^3 + v_h^3}{v_{\mathrm{r}}^3 - v_{\mathrm{ci}}^3} P_{\mathrm{WT}}^{\mathrm{rated}} & v_{\mathrm{ci}} \le v_h < v_{\mathrm{r}} \\ P_{\mathrm{WT}}^{\mathrm{rated}} & v_{\mathrm{r}} \le v_h \le v_{\mathrm{co}} \\ 0 & v_h > v_{\mathrm{co}} \end{cases} \tag{5.26}$$

where v_{ci}, v_{r} and v_{co} represent the cut-in, rated, and cut-out wind speeds, respectively. v_h represents the hourly wind speed.

(2). *Solar energy*: In this study, solar collectors are employed to capture the solar energy and transfer it to heat water. The hourly actual power available from the solar collectors is given by

$$P_{\mathrm{SC}}^{h} = \eta_{\mathrm{SC}} r_h P_{\mathrm{SC}}^{\mathrm{rated}} \tag{5.27}$$

where η_{SC} denotes the efficiency of the solar collectors. r_h denotes hourly solar irradiation.

5.2.2 *Multi-objective Interval Optimization Model*

As for the integrated energy system, the integration of wind and solar energies brings about uncertainty to the system. Generally, the higher the penetration of renewable energyies is, the higher the level of uncertainty will be. To reduce the level of uncertainty, forecasting the accurate variation of the renewable resource is important but that is sometimes difficult because of its strong irregularity especially the wind speed (Ai et al. 2013). However, as illustrated in Figs. 5.2 and 5.3, the wind speed and the solar irradiation can be easily predicted by their lower and upper bounds (Zhou et al. 2014; Lorenz et al. 2009). Therefore, the wind speed v_h and the solar irradiation r_h can be formulated as follows:

$$v_h \in [v_h^{\mathrm{L}}, v_h^{\mathrm{R}}] \tag{5.28}$$

$$r_h \in [r_h^{\mathrm{L}}, r_h^{\mathrm{R}}] \tag{5.29}$$

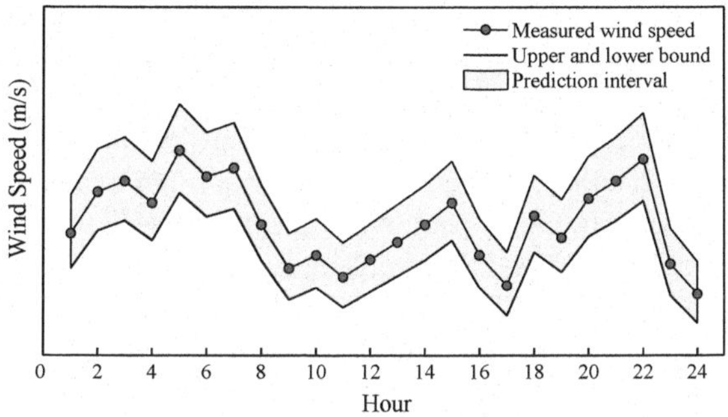

Fig. 5.2 Prediction interval of wind speed

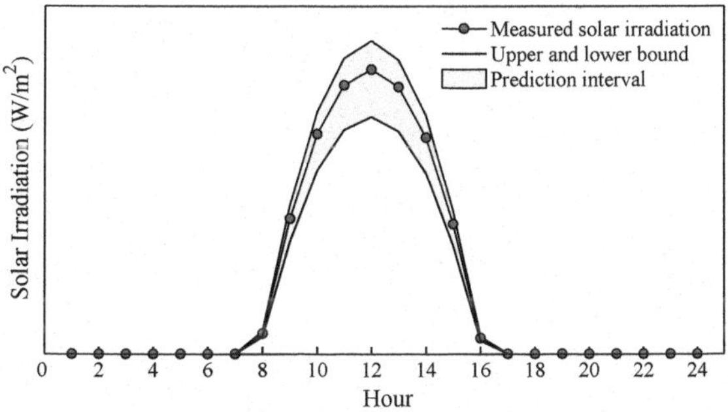

Fig. 5.3 Prediction interval of solar irradiance

where v_h^{L} and v_h^{R} denote the lower and upper bounds of the hourly wind speed, respectively. r_h^{L} and r_h^{R} denote the lower and upper bounds of the hourly solar irradiation, respectively.

As the uncertain variables are tackled as interval variables, the problem shown in (5.3) is converted into an interval optimization problem with both linear and nonlinear constraints. To well present the method, we write the problem (5.3) with interval variables in its general form as follows:

$$\begin{aligned} \min \ & C(X, Y, U) \\ \text{s.t.} \ & g(X, Y, U) = 0 \\ & h(X, Y, U) \leq 0 \\ & U \in [U^{\mathrm{L}}, U^{\mathrm{R}}] \end{aligned} \tag{5.30}$$

where C denotes the objective function. X and Y represent the vectors of independent and dependent variables as shown in (5.1) and (5.2) in Sect. 5.2.1.2, respectively. U represents the vector of interval variables including the hourly wind speed v_h and solar irradiation r_h, which are bounded by the lower bound U^{L} and the upper bound U^{R}. g and h denote the equality and inequality constraints as described in Sect. 5.2.1.2, respectively.

It can be seen that $C(X, Y, U)$ is now an interval number, which is resulted by the interval variables U. Therefore, the method for optimizing the interval objective is ctritical to the result. In the recent decades, a few studies have been conducted on developing various interval optimization technologies. However, most of those studies merely optimized the worst case (Zhou et al. 2014) or the average of objective (Wang et al. 2011), which failed to consider the risk caused by the uncertainties (Li et al. 2014a, b). Moreover, the optimal solution obtained by such kind of methods is proved to be highly sensitive to the interval of the uncertain variables (Wu et al. 2012). In order to address this problem, Jiang et al. (2008) proposed an interval optimization model which took both the mean and risk into consideration by simultaneously minimizing the average and deviation of the objective. First, in this method, the average and deviation of the objective are, respectively, formulated as the midpoint and radius of the objective interval as follows:

$$\begin{aligned} m(C(X,\ Y,\ U)) &= \frac{1}{2}(C^{\mathrm{L}}(X) + C^{\mathrm{R}}(X)) \\ \tau(C(X,\ Y,\ U)) &= \frac{1}{2}(C^{\mathrm{R}}(X) - C^{\mathrm{L}}(X)) \end{aligned} \tag{5.31}$$

where m and τ denote the value of the midpoint and radius of the objective interval, respectively. $C^{\mathrm{L}}(X)$ and $C^{\mathrm{R}}(X)$ denote the lower and upper bounds of the objective interval corresponding to a specific X, respectively, both of which can be simply obtained by solving the following suboptimization problems:

$$\begin{aligned} C^{\mathrm{L}}(X) &= \min_{U}\ C(X,\ Y,\ U) \\ C^{\mathrm{R}}(X) &= \max_{U}\ C(X,\ Y,\ U) \end{aligned} \tag{5.32}$$

As soon as the lower and upper bounds of the objective interval are available, the average and deviation of it can be calculated through (5.31).

In interval method, for a minimization problem, an interval number, C_1, is said to be better than another, C_2, if both the midpoint and radius of C_1 are smaller than those of C_2, i.e., $m(C_1) \leq m(C_2)$ and $\tau(C_1) \leq \tau(C_2)$. Therefore, the midpoint and the radius of the objective should be minimized simultaneously, which can be formulated as a multi-objective problem. In Jiang's method (Jiang et al. 2008), a weighting factor is used to transfer the two objectives into a single-objective problem. It is no doubt that the weighting factor could reflect the preference of the investor. Nevertheless, it is sometimes difficult to judge the risk aversion attitude of the investor. Hence, in this research, the average and deviation of the objective are directly treated as two conflict objectives to obtain a trade-off relationship between them, by which the weighting factor is no longer needed. By substituting (5.32) into (5.31), the multi-objective interval optimization model can be formulated as follows:

$$
\begin{aligned}
&\min\,[m(C(X,\ Y,\ U)),\ \tau(C(X,\ Y,\ U))]\\
\text{s.t. } & m(C(X,\ Y,\ U)) = \frac{1}{2}(\min_{U}\ C(X,\ Y,\ U) + \max_{U}\ C(X,\ Y,\ U))\\
& \tau(C(X,\ Y,\ U)) = \frac{1}{2}(\max_{U}\ C(X,\ Y,\ U) - \min_{U}\ C(X,\ Y,\ U))\\
& g(X, Y, U)\ = 0\\
& h(X, Y, U)\ \le 0\\
& U \in [U^{\mathrm{L}}, U^{\mathrm{R}}]
\end{aligned}
\tag{5.33}
$$

By the formulation of (5.33), the uncertain single-objective optimization problem is now transformed into a deterministic multi-objective problem. Furthermore, the problem (5.33) is actually a two-stage problem (Jiang et al. 2008) containing a first-stage multi-objective optimization problem, i.e., simultaneously optimizing m and τ, and two second-stage nonlinear optimization problems, i.e., obtaining the lower and upper bounds of the objective interval. A multi-objective optimization algorithm used to solve the two-stage problem (5.33) is developed, which will be discussed in the next section.

5.2.3 *Multi-objective Optimization Algorithm and Decision-Making Method*

The comprehensive framework including the optimization model and decision-making method for solving the unit sizing problem of the integrated energy system is illustrated in Fig. 5.4. As has been discussed in Sect. 5.2.2, the multi-objective interval optimization model is first developed. Then, the model is solved by using the multi-objective group search optimizer with adaptive covariance and chaotic search, MGSOACC, to get Pareto-optimal solutions. Finally, the optimal unit sizing solution is determined by the evidential reasoning (ER) approach. The detailed introduction of the MGSOACC and ER approach will be discussed in this section.

5.2.4 *Simulation Studies*

5.2.4.1 Parameter Setting

In this section, the proposed comprehensive framework is tested on both the DH system and the DHC system. The performance of the proposed MGSOACC is compared with the non-dominated sorting based algorithm NSGA-II as well as the animal searching behavior based algorithm GSOMP (Wu et al. 2008; Guo et al. 2012). Each of the three algorithms is evaluated in 30 independent runs with 15,000 function evaluations in each run. The numbers of Pareto-optimal solutions for MGSOACC, GSOMP and NSGA-II are all set to be 7.

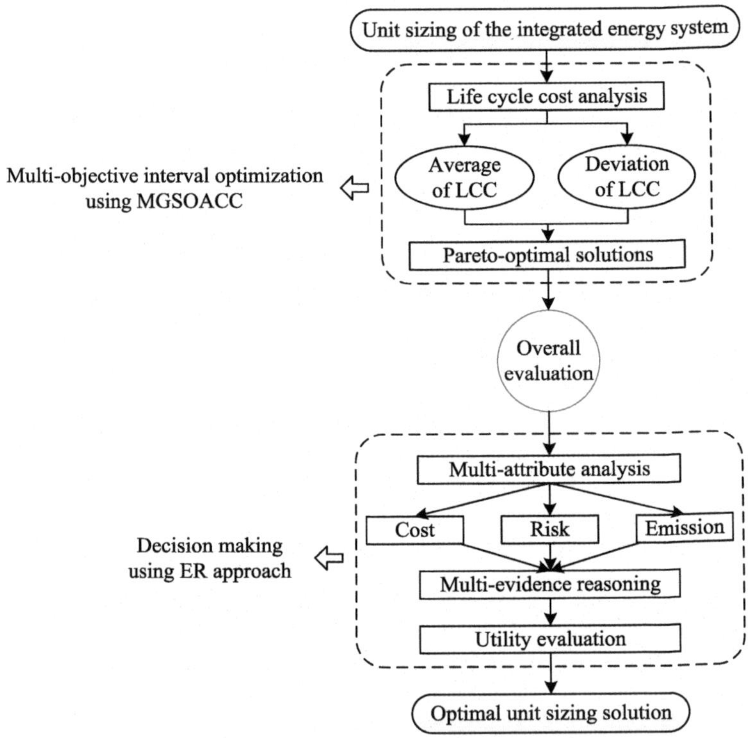

Fig. 5.4 The framework of the multi-objective interval optimization and decision-making

During the planning stage of the integrated energy system, reasonable assumptions and simplifications on its long-term operation are essential. With regard to the hourly variation of wind speed, it seems not sensitive to the clock time because of its strong irregularity. Contrarily, as shown in Fig. 5.3, it is evident that the hourly solar irradiation variates significantly with the clock time. For instance, the solar irradiation during the midday is usually much stronger than that in any other period in a day. In order to compromise between the characteristic of the variation of daily solar irradiation and the computational cost, the peak–low–even method which divides the daily heating and cooling loads into three representative periods is used to stand for the daily operation in this study (Tan et al. 2014). Table 5.1 displays the energy price, as well as the heating and cooling loads during the peak, even, and low periods (Mao 2007). It should be noted that according to the human thermal comfort study (Mao 2007), the peak heating load of residential consumer usually occurs during when the electricity tariff is low while the low-heating load occurs during when the electricity tariff is even. Table 5.2 shows the forecasted intervals of wind speed and solar irradiation during the corresponding hours, which are assumed to be valid for both the heating and cooling seasons.

Table 5.1 Demand and energy price during peak, even, and low periods

Item	Unit	Peak hours	Even hours	Low hours
		18:00–24:00	08:00–18:00	00:00–08:00
Electricity tariff	\$/kWh	0.133	0.079	0.027
Natural gas price	$\$/m^3$	0.482	0.482	0.482
Heating load	kW	59.87	49.79	69.95
Cooling load	kW	93.75	64.62	25.04

Table 5.2 Forcasted interval of wind speed and solar irradiation

Item	Unit	Daytime	Midday	Night
		06:00–10:00 15:00–18:00	10:00–15:00	00:00–06:00 18:00–24:00
Solar irradiation	W/m^2	[150 300]	[375 654]	[0 0]
Wind speed	m/s	[5 14]		

Table 5.3 Pollutant emission coefficient and cost factor

Item	Unit	Pollutant			
		CO	CO_2	NO_x	SO_2
Electricity	g/kWh	0.1083	623	2.88	6.48
Natural gas	g/kWh	0.1702	184.0829	0.6188	0.000928
Environmental value	\$/kg	0.145	0.004125	1.25	0.875

The economic and technique parameters of the DH system and the DHC system are given in Table 5.4. In addition, for the wind turbine, the cut-in, cut-out, and rated wind speeds are set as 3 m/s, 25 m/s and 10 m/s, respectively. It is assumed that the subsidy of the wind energy system is 30% of its initial capital cost (Jiang et al. 2014). The discount rate is set to be 5% in this study. The heating season and the cooling season are assumed to be 5 months and 4 months per year, respectively. Suppose two labors are employed to work in the DH/DHC station and the annual salary is set to be 8026\$ for each labor (Dong and Xu 2015). Table 5.3 shows the pollutant emission factors and the environmental values of electricity and natural gas.

5.2.4.2 Case I: DH System

Figure 5.5 shows the Pareto-optimal solutions obtained by MGSOACC, NSGA-II and GSOMP, respectively, which proves that the MGSOACC outperforms GSOMP and NSGA-II in terms of searching for more evenly distributed and better converged Pareto-optimal solutions. On the other hand, the spread over range of MGSOACC is $[393, 198,\ \ 78, 230] \times [411, 237,\ \ 42, 412]$, which is wider than

Table 5.4 The economic and technique parameter settings of DH and DHC systems

Item	Unit	Energy supply unit					
		WT	SC	EB	GB	RC	AC
Capacity dependent cost	\$/kW (\$/m^2)	1123	225	67.88	100.5	80	105
Capacity independent cost	\$	–	4000	–	–	–	–
Installation cost coefficient	–	10%	10%	10%	10%	10%	10%
Fixed maintenance cost coefficient	–	2%	2%	4%	5%	3%	4%
Variable maintenance cost	\$/kWh	–	–	0.006	0.004	–	–
Life time	year	20	20	20	20	20	20
Efficiency	–	–	78.8%	95%	90%	–	–
Coefficient of performance	–	–	–	–	–	4	0.7

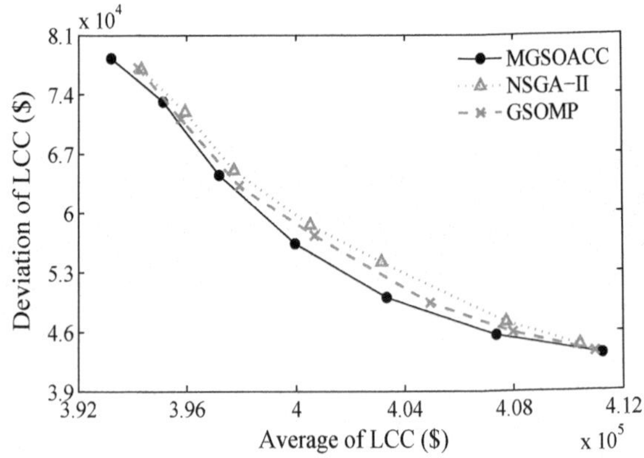

Fig. 5.5 Pareto-optimal solutions obtained by MGSOACC, NSGA-II and GSOMP of DH system

that of NSGA-II, $[394,315,\ 76,934] \times [410,415,\ 44,428]$, and that of GSOMP, $[394,181,\ 77,138] \times [410,963,\ 43,657]$.

Figure 5.6 shows the bar chart for the average and deviation of LCC of the obtained 7 solutions which are denoted as $s_1,\ s_2,\ \ldots,\ s_7$. It is evident that the average of LCC will increase while the deviation of it decreases, which demonstrates that the average of the objective is conflict to the deviation. Therefore, there is no doubt that to optimize the average and deviation of the objective simultaneously is necessary. To further demonstrate the necessity, the obtained 7 Pareto-optimal solutions are compared with the solution obtained by traditional single-objective interval optimization which only the averaged LCC is optimized. The corresponding objective values of the 8 solutions are displayed in Table 5.5. Noted that s_0 is a single-objective optimization solution obtained by GSO.

It can be seen that if s_0 is accepted by the investor, the lowest investment cost (average) will be achieved. However, the investor has to take on a high risk (deviation), which is about 20% of the total amount of investment. Although a tempting cost can

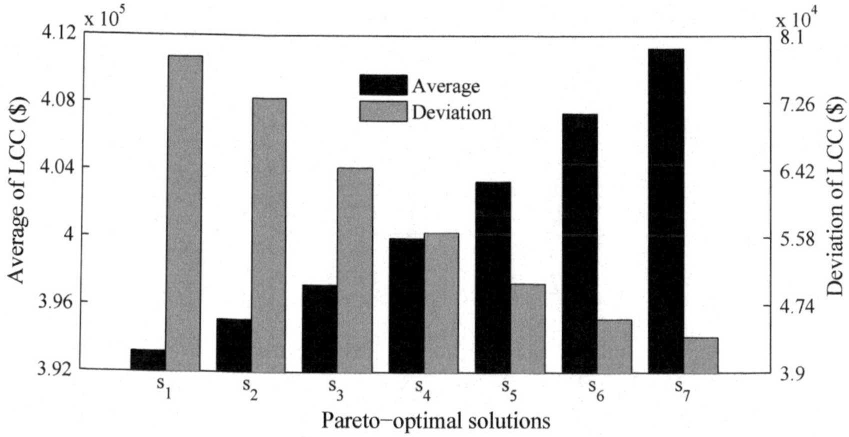

Fig. 5.6 Average and deviation of LCC as for $s_1, s_2, \ldots, s_7$

Table 5.5 Cost, risk, and emission values as for $s_0, s_1, \ldots, s_7$

Solution	Average of LCC ($) (Cost)	Deviation of LCC ($) (Risk)	Total pollutant emission (kg) (Emission)
s_0	**392, 488**	78, 399	1, 165, 554
s_1	393, 198	78, 230	**1, 163, 097**
s_2	395, 114	73, 108	1, 250, 414
s_3	397, 168	64, 483	1, 287, 917
s_4	399, 948	56, 383	1, 387, 194
s_5	403, 322	50, 063	1, 456, 467
s_6	407, 363	45, 634	1, 548, 158
s_7	411, 237	**43, 412**	1, 684, 210

be achieved by s_0, it will result in an unbearable risk to a business investment unless the investor is aggressive enough. On the contrary, if s_7 is accepted, the investment risk will be reduced to 10.56%, which means that the s_7 is not sensitive to the uncertain environment of wind and solar energies. However, the averaged LCC of s_7 is increased inevitably because it prefers a low risk without considering the increased cost. Therefore, both the unit sizing solution s_0 and s_7 are not advisable for the investor considering the economic performance.

The multi-objective interval optimization model provides different levels of cost and risk obtained by achieving various unit sizing solutions under the uncertain environment of wind and solar energies, by which the investor is able to trade off between the conflict interests of cost and risk. In addition to that, pollutant emissions indicating the environmental performance of the system should also be considered at the decision-making stage. Therefore, to make a compromise between "cost",

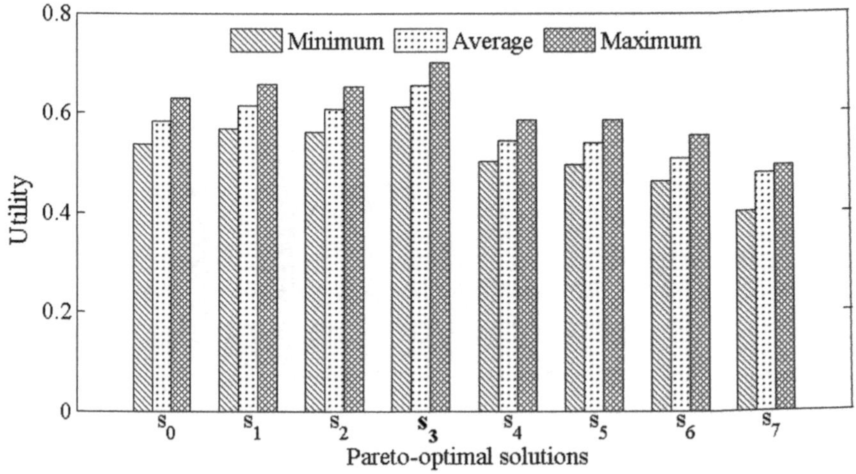

Fig. 5.7 Minimum, average, and maximum utilities as for $s_1, s_2, \ldots, s_7$

"risk", and "emission" as shown in Table 5.5, ER approach is applied to make the final decision.

By ranking the numerical magnitudes of the basic attributes, the basic assessments for each Pareto-optimal solution can be assigned, which are presented in Table 5.6. As the economic benefits are the main consideration of a business investment, they should be assigned with greater weights. Therefore, the relative weights of these three basic attributes are set to be $\omega = [0.4,\ 0.4,\ 0.2]$. Additionally, the evaluation grades are set to be $H = \{$Poor, Unsatisfied, Normal, Good, Excellent, Uncertain$\}$. It should be noted that to avoid loss of generality, the unassigned assessment for each attribute is set to be 0.1.

By applying the evidence combination method of ER approach, the distributed assessments can be obtained, the results of which are displayed in Table 5.7. However, it is difficult to distinguish the pros and cons of the solutions merely from the distributed assessments. Thus, the utility functions are applied to rank these solutions. Figure 5.7 shows the minimum, average, and maximum utilities for each solution. It can be seen from Fig. 5.7 that the ranking of the 8 solutions is $s_3 \succ s_1 \succ s_2 \succ s_0 \succ s_4 \succ s_5 \succ s_6 \succ s_7$. Therefore, s_3 is selected as the final unit sizing solution for the DH system as it gains the most average utility among the 8 solutions.

Furthermore, to verify the superiority of s_3, it is compared with the unit sizing solution s_8 proposed in our previous work (Jiang et al. 2014). Table 5.8 displays the two unit sizing solutions and their corresponding objective values. It is clear that the total installed capacity of renewable energy units of s_3 is significantly larger than that of s_8. Though adopting s_3 inevitably increases the initial investment, the averaged LCC of it will be reduced because the high renewable utilization ratio of s_3 achieves a low operation cost during the life cycle of the DH system. Moreover, it can be seen

Table 5.6 Basic assessments for $s_0, s_1, \ldots, s_7$

Solution	Attribute	Evaluation grade				
		Poor	Unsatisfied	Normal	Good	Excellent
s_0	Cost	0	0	0	0.2	0.7
	Risk	0.8	0.1	0	0	0
	Emission	0	0	0.1	0.2	0.6
s_1	Cost	0	0	0	0.3	0.6
	Risk	0.6	0.3	0	0	0
	Emission	0	0	0	0.2	0.7
s_2	Cost	0	0	0.2	0.2	0.5
	Risk	0.4	0.3	0.2	0	0
	Emission	0	0	0.1	0.4	0.4
s_3	Cost	0	0	0.2	0.3	0.4
	Risk	0.1	0.2	0.6	0	0
	Emission	0	0	0.3	0.3	0.3
s_4	Cost	0.1	0.3	0.4	0.1	0
	Risk	0	0.1	0.7	0.1	0
	Emission	0	0	0.4	0.4	0.1
s_5	Cost	0.2	0.4	0.3	0	0
	Risk	0	0	0.4	0.4	0.1
	Emission	0	0.2	0.5	0.2	0
s_6	Cost	0.5	0.4	0	0	0
	Risk	0	0	0	0.3	0.6
	Emission	0.2	0.4	0.3	0	0
s_7	Cost	0.7	0.2	0	0	0
	Risk	0	0	0	0.1	0.8
	Emission	0.6	0.3	0	0	0

Table 5.7 Distributed assessments as for $s_0, s_1, \ldots, s_7$

Solution	Poor	Unsatisfied	Normal	Good	Excellent	Uncertain
s_0	0.3211	0.0401	0.0157	0.1155	0.4161	0.0915
s_1	0.2404	0.1202	0.0000	0.1573	0.3908	0.0914
s_2	0.1611	0.1208	0.1911	0.1513	0.2839	0.0918
s_3	0.0389	0.0778	0.4110	0.1709	0.2126	0.0888
s_4	0.0367	0.1537	0.5711	0.1404	0.0143	0.0837
s_5	0.0773	0.1923	0.4114	0.1923	0.0386	0.0881
s_6	0.2466	0.2436	0.0479	0.1228	0.2456	0.0934
s_7	0.4152	0.1329	0.0000	0.0401	0.3205	0.0914

Table 5.8 Unit sizing solutions and objective values as for s_3 and s_8

Solution	Facility				LCC	
	WT (kW)	SC (m^2)	GB (kW)	EB (kW)	Average ($)	Deviation ($)
s_3	23	68	38	41	397, 168	64, 483
s_8	20	15	45	40	401, 676	60, 295

Table 5.9 Pollutant emissions as for s_3 and s_8

Solution	Pollutant				Total pollutant emission (kg)
	CO (kg)	CO_2 (kg)	NO_x (kg)	SO_2 (kg)	
s_3	431	1, 271, 649	5, 525	10, 312	1, 287, 917
s_8	539	1, 403, 610	5, 991	10, 504	1, 420, 646

from Table 5.9 that in comparison with s_3, insufficient sizing of renewable units of s_8 leads to a heavier dependence on conventional energies and ultimately, a higher pollutant emission.

It should be emphasized that the attributes considered in the decision-making contains not only economic criteria but also environmental criterion. On one hand, according to the return and risk relation in portfolio selection proposed by Markowitz (1952), the investor has to take on additional risk to gain more return, or reduce risk by giving up expected profit, which demonstrates that there must be a compromise between the two economic criteria. Therefore, although s_8 gains a less deviation indicating a lower risk, adopting it means more payment should be afforded by the investor. On the other hand, as has been discussed above, inadequate sizing of wind and solar energy units of s_8 results in a huge increase in the amount of total pollutant emission, which is up to about 1.33×10^5 kg. However, it is significantly not worthwhile for a rational investor to covet a little reward by adopting s_8 at such a high price in terms of deterioration in both cost and emission. Hence, s_3 is proved to be a superior solution based on the comprehensive consideration of cost, risk, and emission.

5.2.4.3 Case II: DHC System

For the DHC system, the Pareto-optimal solutions obtained by MGSOACC, NSGA-II, and GSOMP are shown in Fig. 5.8. Better converged and more evenly distributed Pareto-optimal solutions obtained by MGSOACC than those obtianed by NSGA-II and GSOMP can be observed from the figure. Moreover, a wider spread range of Pareto-optimal solutions is obtained by MGSOACC, [478, 098, 106, 668] × [491, 281, 72, 509], than that of either NSGA-II, [478, 366, 104, 451] × [491, 114, 72, 854], or GSOMP, [478, 484, 105, 971] × [491, 004, 74, 858].

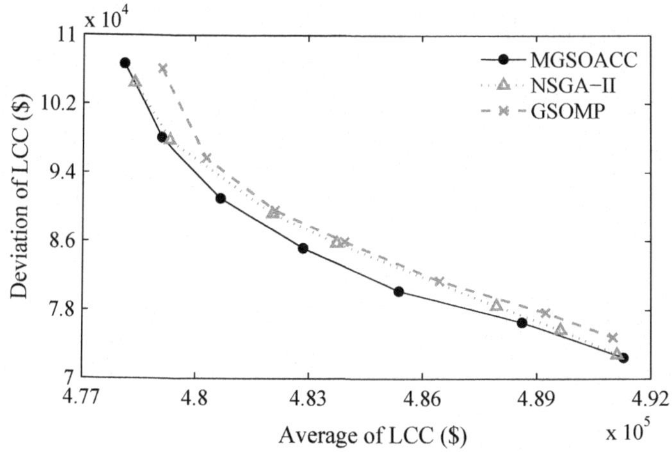

Fig. 5.8 Pareto-optimal solutions obtained by MGSOACC, NSGA-II and GSOMP of DHC system

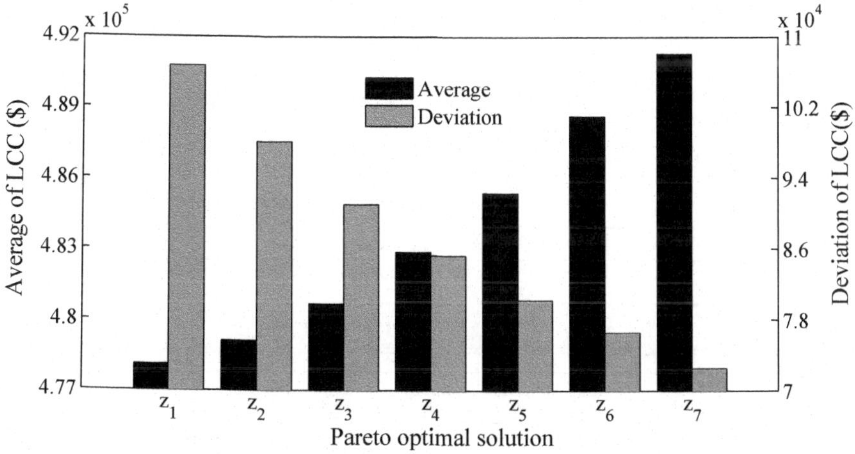

Fig. 5.9 Average and deviation of LCC as for $z_1, z_2, \ldots, z_7$

The average and deviation of LCC of the obtained optimal unit sizing solutions are illustrated in Fig. 5.9, in which $z_1, z_2, \ldots, z_7$ represent the 7 solutions, respectively. It is obvious that the less average of LCC the unit sizing solution gains, the more deviation it takes on. As a solution that obtains the minimum average and the least deviation does not exist, it is necessary to take both of them into consideration.

The corresponding objective values of the 7 Pareto-optimal solutions are displayed in Table 5.10. It is noted that z_0 is the unit sizing solution obtained by GSO with the only consideration of averaged LCC. It can be seen from the table that z_0 achieves the least cost at the price of high risk which equals 22.55% of the cost. Such a high

Table 5.10 Cost, risk, and emission values as for $z_0, z_1, \ldots, z_7$

Solution	Average of LCC ($) (Cost)	Deviation of LCC ($) (Risk)	Total pollutant emission (kg) (Emission)
z_0	**476, 843**	107, 360	1, 584, 554
z_1	478, 098	106, 668	1, 615, 891
z_2	479, 102	98, 078	1, 627, 973
z_3	480, 670	91, 018	**1, 536, 278**
z_4	482, 850	85, 211	1, 618, 716
z_5	485, 372	80, 224	1, 689, 910
z_6	488, 622	76, 575	1, 760, 053
z_7	491, 281	**72, 509**	1, 826, 498

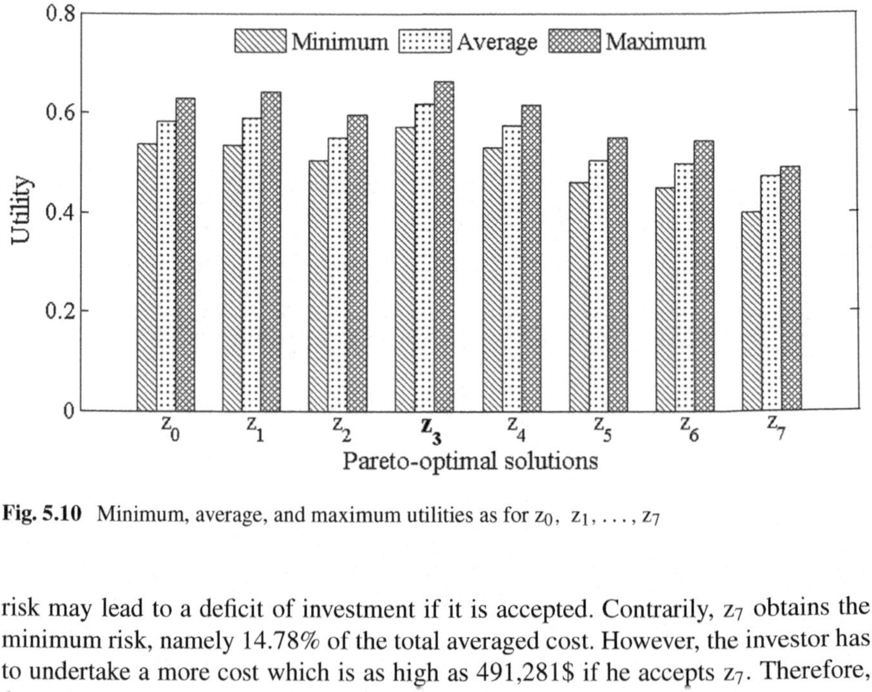

Fig. 5.10 Minimum, average, and maximum utilities as for $z_0, z_1, \ldots, z_7$

risk may lead to a deficit of investment if it is accepted. Contrarily, z_7 obtains the minimum risk, namely 14.78% of the total averaged cost. However, the investor has to undertake a more cost which is as high as 491,281$ if he accepts z_7. Therefore, for a shrewd businessman, z_0 and z_7 are undesirable.

As the same approach as that in Case I, ER approach is applied to make the final decision in order to conduct a trade-off solution between cost, risk, and emission. The relative weights ω_i are also set to be $\omega = [0.4,\ 0.4,\ 0.2]$ in this case. Tables 5.11 and 5.12 display the basic and distributed assessments for each Pareto-optimal solution, respectively. Figure 5.10 illustrates the utility evaluation results of all the solutions. It is evident that the ranking of the 8 solutions is $z_3 \succ z_1 \succ z_0 \succ z_4 \succ z_2 \succ z_5 \succ z_6 \succ z_7$. Therefore, through the ER approach, z_3 is finally determined as the unit sizing solution of DHC system.

Table 5.11 Basic assessments for $z_0, z_1, \ldots, z_7$

Solution	Attribute	Evaluation grade				
		Poor	Unsatisfied	Normal	Good	Excellent
z_0	Cost	0	0	0	0.2	0.7
	Risk	0.8	0.1	0	0	0
	Emission	0	0	0	0.4	0.5
z_1	Cost	0	0	0.1	0.2	0.6
	Risk	0.5	0.4	0	0	0
	Emission	0	0	0.1	0.3	0.4
z_2	Cost	0	0	0.2	0.3	0.4
	Risk	0.3	0.5	0.1	0	0
	Emission	0	0.2	0.5	0.2	0
z_3	Cost	0	0	0.4	0.4	0.1
	Risk	0.2	0.4	0.2	0.1	0
	Emission	0	0	0	0.1	0.8
z_4	Cost	0	0.4	0.3	0.2	0
	Risk	0	0.3	0.3	0.3	0
	Emission	0	0	0.5	0.2	0.2
z_5	Cost	0.3	0.3	0.3	0	0
	Risk	0	0	0.3	0.4	0.2
	Emission	0.2	0.4	0.3	0	0
z_6	Cost	0.4	0.4	0.1	0	0
	Risk	0	0	0	0.35	0.55
	Emission	0.5	0.4	0	0	0
z_7	Cost	0.7	0.2	0	0	0
	Risk	0	0	0	0.15	0.85
	Emission	0.6	0.3	0	0	0

Table 5.12 Distributed assessments as for $z_0, z_1, \ldots, z_7$

Solution	Poor	Unsatisfied	Normal	Good	Excellent	Uncertain
z_0	0.3220	0.0403	0.0000	0.1512	0.3947	0.0918
z_1	0.2050	0.1640	0.0576	0.1347	0.3319	0.1069
z_2	0.1215	0.2440	0.2210	0.1591	0.1620	0.0924
z_3	0.0804	0.1609	0.2614	0.2320	0.1736	0.0917
z_4	0.0000	0.2920	0.3509	0.2418	0.0294	0.0859
z_5	0.1554	0.1920	0.3250	0.1583	0.0791	0.0902
z_6	0.2610	0.2412	0.0405	0.1419	0.2229	0.0924
z_7	0.4152	0.1329	0.0000	0.0601	0.3004	0.0914

Table 5.13 Unit sizing solutions and objective values of z_3 and z_8

Solution	Facility						LCC	
	WT (kW)	SC (m^2)	GB (kW)	EB (kW)	RC (kW)	AC (kW)	Average ($)	Deviation ($)
z_3	31	119	31	41	82	46	480, 670	91, 018
z_8	20	15	45	40	95	21	488, 827	85, 124

Table 5.14 Pollutant emissions of z_3 and z_8

Solution	Pollutant				Total emission (kg)
	CO (kg)	CO_2 (kg)	NO_x (kg)	SO_2 (kg)	
z_3	564	1, 517, 579	6, 511	11, 624	1,536, 278
z_8	755	1, 942, 482	8, 278	14, 423	1,965, 938

Then, z_3 is compared with the unit sizing solution z_8, which is proposed in Jing et al. (2014). The comparison results of the unit sizing solutions z_3 and z_8 are shown in Table 5.13. Recall that the decision is made upon the synthetical consideration of investment cost and risk, as well as the total pollutant emission. It can be seen that the deviation of z_3 is higher than that of z_8, which means a better adjustment of z_8 to all the uncertain renewable energies. However, such an improvement is achieved at the price of a great waste of renewable sources caused by insufficient total installed capacity of wind and solar energy units of z_8, which is even less than a quarter of that of z_3. Reducing renewable utilization means an increasing burden on fossil fuel energy, especially for the DHC system which operates both in the summer and winter time, because the energy balance constraints should always be satisfied at any time slot. As a result, a higher operation cost and ultimately a higher averaged cost, with even more pollutant emission is inevitably incurred if z_8 is adopted. Furthermore, reducing the utilization of renewable energy violates the intention of developing green and clean energy system as we are now pursuing. Hence, it is not advisable for the investor to adopt z_8 considering the above factors. Alternatively, z_3 is, therefore, superior to z_8 in terms of satisfying both economic and environmental requirements (Table 5.14).

In this section, the multi-objective interval optimization model has been presented to solve the unit sizing problem of small-scale integrated energy systems considering the integration of uncertain wind and solar energies. This model takes into account both the cost and risk of business investment in the integrated energy system through, respectively, manifesting them by the average and deviation of LCC, in the uncertain environment of wind and solar energies. By converting the uncertain single-objective problem into a deterministic multi-objective problem, the competing relationship between cost saving and risk reduction are simultaneously considered. The result obtained based on either of the DH and DHC system has verified the necessity of considering both the cost and risk while solving the unit sizing problem of integrated energy systems.

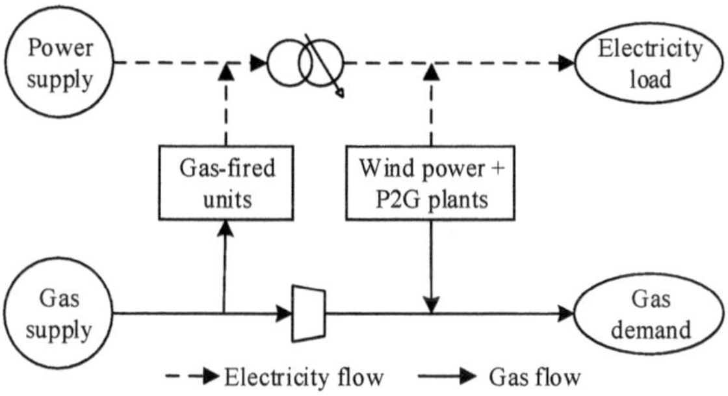

Fig. 5.11 A framework of electricity-gas systems interconnected with gas-fired units and P2G plants

In addition, to address the problem, the multi-objective group search optimizer with adaptive covariance and chaotic search is developed by employing chaotic search to enhance the global searching capability. Comparing MGSOACC with GSOMP and NSGA-II, the more evenly distributed Pareto-optimal solutions have demonstrated the searchability of MGSOACC.

Furthermore, a multi-attribute decision-making approach is applied to process the investor's cognition on cost, risk, and emission. Comparison results obtained based on both of the two case studies have demonstrated that the ER approach is efficient in terms of determining a superior unit sizing solution with not only low cost and acceptable risk but also low pollutant emission.

Finally, it is worth mentioning that the framework including optimization and decision-making proposed in this section can be applied to guide the investment behavior of the investor who has to take multiple interests into consideration, in various business investments such as investment in more comprehensive and/or larger scale integrated energy systems with renewable energy integrated.

5.3 Multi-stage Contingency-Constrained Co-planning for Integrated Energy Systems

5.3.1 Formulation of MCC Model

The proposed model of electricity-gas systems interconnected with gas-fired units and P2G plants can be illustrated in Fig. 5.11. The systems include electricity transmission system, gas supply system, gas-fired units, wind power energy, P2G plants, electricity load and gas demand. Wind power resources give priority to electricity load, while the surplus ones will be converted into hydrogen or synthetic natural gas (SNG) via P2G plants to supply the gas demand. Thus, the combination of gas-fired units and P2G plants realize the two-way coupling of electricity system and natural gas system. The detailed formulations of the MCC model can be described as follows.

5.3.1.1 Objective Function

The MCC model considers the long-term co-planning for EGS with the short-term operation constraints. The objective (5.34) of the MCC model is to minimize the total co-planning cost (consisting of investment cost, operation cost, and curtailed wind energy cost under the normal operating condition), while meeting the transmission N-1 contingency constraints.

$$\min C = C^{\text{inv}} + C^{\text{ope}} + C^{\text{cur}} \tag{5.34}$$

$$\begin{aligned} C^{\text{inv}} = & \sum_{t=1}^{N_{\mathrm{T}}} \sum_{i \in CG} \frac{C_i^{\text{gen}} P_i^{\text{gen,rate}} \left(X_{i,t} - X_{i,(t-1)}\right)}{(1+d)^{t-1}} + \sum_{t=1}^{N_{\mathrm{T}}} \sum_{\mathrm{l}_k \in CL} \frac{C_{\mathrm{l}_k}^{\text{line}} \left(Y_{\mathrm{l}_k,t} - Y_{\mathrm{l}_k,(t-1)}\right)}{(1+d)^{t-1}} \\ & + \sum_{t=1}^{N_{\mathrm{T}}} \sum_{\mathrm{p}_j \in CP} \frac{C_{\mathrm{p}_j}^{\text{pipe}} \left(Z_{\mathrm{p}_j,t} - Z_{\mathrm{p}_j,(t-1)}\right)}{(1+d)^{t-1}} + \sum_{t=1}^{N_{\mathrm{T}}} \sum_{w \in CU} \frac{C_w^{\text{P2G}} \left(U_{w,t} - U_{w,(t-1)}\right)}{(1+d)^{t-1}} - C^{\text{sal}} \end{aligned} \tag{5.35}$$

$$\begin{aligned} C^{\text{sal}} = & \sum_{t=1}^{N_{\mathrm{T}}} \left(\sum_{i \in CG} \chi_{i,t} C_i^{\text{gen}} P_i^{\text{gen,rate}} \left(X_{i,t} - X_{i,(t-1)}\right) + \sum_{\mathrm{l}_k \in CL} \chi_{\mathrm{l}_k,t} C_{\mathrm{l}_k}^{\text{line}} \left(Y_{\mathrm{l}_k,t} - Y_{\mathrm{l}_k,(t-1)}\right) \right. \\ & \left. + \sum_{\mathrm{p}_j \in CP} \chi_{p,t} C_{\mathrm{p}_j}^{\text{pipe}} \left(Z_{\mathrm{p}_j,t} - Z_{\mathrm{p}_j,(t-1)}\right) + \sum_{w \in CU} \chi_{w,t} C_w^{\text{P2G}} \left(U_{w,t} - U_{w,(t-1)}\right) \right) \frac{1}{(1+d)^{N_{\mathrm{T}}-1}} \end{aligned} \tag{5.36}$$

$$C^{\text{ope}} = \sum_{t=1}^{N_{\mathrm{T}}} \sum_{e=1}^{N_{\mathrm{E}}} E_e \left(\sum_{s=1}^{N_{\mathrm{S}}} \frac{f_{t,e,s}^{\text{supply}} c_{t,e,s}^{\text{supply}}}{(1+d)^{t-1}} + \sum_{i=1}^{N_{\mathrm{C}}} \frac{P_{t,e,i}^{\text{coal}} c_{t,e,i}^{\text{coal}}}{(1+d)^{t-1}} \right) \tag{5.37}$$

$$C^{\text{cur}} = \sum_{t=1}^{N_{\mathrm{T}}} \sum_{e=1}^{N_{\mathrm{E}}} E_e \left(\sum_{w=1}^{N_{\mathrm{W}}} \frac{P_{t,e,w}^{\text{cur}} c_{t,e,w}^{\text{cur}}}{(1+d)^{t-1}} \right) \tag{5.38}$$

where $CL = \{\mathrm{l}_k \to (m_k, n_k) \,|\, k = 1, 2, \ldots, N_{\mathrm{CL}}\}$. l_k represents the transmission line from bus m_k to n_k. $CP = \left\{\mathrm{p}_j \to \left(a_j, b_j\right) | j = 1, 2, \ldots, N_{\mathrm{CP}}\right\}$. p_j represents the gas pipeline from node a_j to b_j. The first four items in (5.35) indicate investment costs of gas-fired units, transmission lines, gas pipelines, and P2G plants, respectively. The last item C^{sal} is the salvage value of new assets. A higher salvage factor χ indicates a lower depreciation at the end of the planning horizon (5.36). The operation cost (5.37) includes those of gas suppliers and coal units for all selected scenarios $e \in \{1, 2, \ldots, N_{\mathrm{E}}\}$, during each planning stage $t \in \{1, 2, \ldots, N_{\mathrm{T}}\}$. The penalty cost for curtailed wind energy is expressed as (5.38). Note that in this study, two kinds of generator units exist in the initial electricity-gas systems, including gas-fired units and coal units. However, the expansion planning of coal units is not considered due to their large CO_2 emissions.

5.3.1.2 Variables

The decision variables include binary variables X, Y, and Z, denoting the investment states of gas-fired units, transmission lines, and gas pipelines; integer variable U denoting the installed capacity of P2G plants, as well as continuous variables P^{coal}, f^{supply}, and P^{cur}, denoting the generation outputs of coal units, natural gas production, and possible wind energy curtailment, respectively.

5.3.1.3 Constraints

The MCC model is subjected to the following constraints, including sequential investment states of new assets, operational constraints of power system, natural gas fuel supply availability, physical limits of interconnected media, and transmission N-1 contingency constraints. For all constraints, $t \in \{1, 2, ..., N_{\mathrm{T}}\}$, $e \in \{1, 2, ..., N_{\mathrm{E}}\}$.

(a) Sequential investment states of new assets

$$X_{i,(t-1)} \leq X_{i,t} \tag{5.39}$$

$$Y_{\mathrm{l}_k,(t-1)} \leq Y_{\mathrm{l}_k,t} \tag{5.40}$$

$$Z_{\mathrm{p}_j,(t-1)} \leq Z_{\mathrm{p}_j,t} \tag{5.41}$$

Note that once a candidate component is installed, its investment state will change from 0 to 1 for the remaining years.

(b) Operational constraints of power system

$$\boldsymbol{I}^{\text{line}} \boldsymbol{P}^{\text{line}}_{t,e} + \boldsymbol{I}^{\text{gen}} \boldsymbol{P}^{\text{gen}}_{t,e} + \boldsymbol{I}^{\text{coal}} \boldsymbol{P}^{\text{coal}}_{t,e} + \boldsymbol{I}^{\text{wind}} \boldsymbol{P}^{\text{wind}}_{t,e} = \boldsymbol{I}^{\text{load}} \boldsymbol{P}^{\text{load}}_{t,e} \tag{5.42}$$

$$\left| P^{\text{line}}_{t,e,\mathrm{l}_k} + y_{\mathrm{l}_k} \left(\theta_{t,e,m} - \theta_{t,e,n}\right) \right| \leq M \left(1 - Y_{\mathrm{l}_k,t}\right), \quad \forall \mathrm{l}_k \in CL \tag{5.43}$$

$$\left| P^{\text{line}}_{t,e,\mathrm{l}_k} \right| \leq P^{\text{line,max}}_{\mathrm{l}_k} Y_{\mathrm{l}_k,t}, \quad \forall \mathrm{l}_k \in CL \tag{5.44}$$

$$P^{\text{coal,min}}_{i} \leq P^{\text{coal}}_{t,e,i} \leq P^{\text{coal,max}}_{i}, \quad \forall i \in EC \tag{5.45}$$

$$P^{\text{wind}}_{t,e,w} + P^{\text{P2G}}_{t,e,w} + P^{\text{cur}}_{t,e,w} = P^{\text{wind,pre}}_{t,e,w} \tag{5.46}$$

$$\theta_{ref} = 0 \tag{5.47}$$

The power balance is given in (5.42). The power flow in a new topology structure is limited by (5.43)–(5.44). If a candidate line is not installed, Eq. (5.43) is relaxed and (5.44) sets the flow to zero. The available energy outputs of coal units are limited by the physical characteristics (5.45). The wind power balance formulation of each wind farm is described as (5.46). The angle of slack bus is set to zero (5.47).

(c) Natural gas fuel supply availability

$$\boldsymbol{K}^{\text{pipe}}\boldsymbol{f}_{t,e}^{\text{pipe}} + \boldsymbol{K}^{\text{supply}}\boldsymbol{f}_{t,e}^{\text{supply}} + \boldsymbol{K}^{\text{P2G}}\boldsymbol{f}_{t,e}^{\text{P2G}} = \boldsymbol{K}^{\text{gen}}\boldsymbol{f}_{t,e}^{\text{gen}} + \boldsymbol{K}^{\text{com}}\boldsymbol{f}_{t,e}^{\text{com}} + \boldsymbol{K}^{\text{load}}\boldsymbol{f}_{t,e}^{\text{load}} \quad (5.48)$$

$$f_{t,e,\mathrm{p}_j}^{\text{com}} = k^{\text{com}} f_{t,e,\mathrm{p}_j}^{\text{pipe}}, \quad \forall \mathrm{p}_j \in EP \quad (5.49)$$

$$f_{\mathrm{p}_j}^{\text{pipe,min}} \le f_{t,e,\mathrm{p}_j}^{\text{pipe}} \le f_{\mathrm{p}_j}^{\text{pipe,max}}, \quad \forall \mathrm{p}_j \in EP \quad (5.50)$$

$$f_{\mathrm{p}_j}^{\text{pipe,min}} Z_{\mathrm{p}_j,t} \le f_{t,e,\mathrm{p}_j}^{\text{pipe}} \le f_{\mathrm{p}_j}^{\text{pipe,max}} Z_{\mathrm{p}_j,t}, \quad \forall \mathrm{p}_j \in CP \quad (5.51)$$

$$f_s^{\text{supply,min}} \le f_{t,e,s}^{\text{supply}} \le f_s^{\text{supply,max}} \quad (5.52)$$

where $EP = \{\mathrm{p}_j \to (a_j, b_j) \,|\, j = 1, 2, \ldots, N_{\text{EP}}\}$. The gas flow is a function of nodal pressure and pipeline characteristics, which is generally solved by Newton method. However, to accelerate the process, we use the simplified transportation model in this long-term planning horizon. The node balance is given in (5.48). According to Borraz-Sánchez and Ríos-Mercado (2009), the gas consumption of compressor stations accounts for typically 3–5% of the transported gas. Therefore, the expression of gas compressor is simplified as a coefficient k^{com} multiplied by transported gas (5.49). Pipeline gas flow is limited by the upper bound (5.50)–(5.51). The available gas production is limited by the physical characteristics of the gas source (5.52).

(d) Physical limits of interconnected media

$$f_{t,e,g}^{\text{gen}} = k^{\text{gen}} P_{t,e,i}^{\text{gen}}, \quad \forall i \in \{EG, CG\} \quad (5.53)$$

$$P_i^{\text{gen,min}} \le P_{t,e,i}^{\text{gen}} \le P_i^{\text{gen,max}}, \quad \forall i \in EG \quad (5.54)$$

$$P_i^{\text{gen,min}} X_{i,t} \le P_{t,e,i}^{\text{gen}} \le P_i^{\text{gen,max}} X_{i,t}, \quad \forall i \in CG \quad (5.55)$$

$$f_{t,e,w}^{\text{P2G}} = k^{\text{P2G}} P_{t,e,w}^{\text{P2G}} \quad (5.56)$$

$$0 \le P_{t,e,w}^{\text{P2G}} \le U_{w,t} \quad (5.57)$$

The natural gas required by each gas-fired unit depends on its hourly dispatch (5.53). The available energy outputs of existing and candidate gas-fired units are limited by the physical characteristics (5.54)–(5.55). The energy conversion equation of P2G plants is written as (5.56). The available energy input of P2G plants is limited by the installed capacity (5.57).

(e) Transmission N-1 contingency constraints

$$\left|P^{\text{line}}_{t,e,\text{l}_k} + y_{\text{l}_k}\left(\theta_{t,e,m} - \theta_{t,e,n}\right)\right| \leq M\left(1 - D_{\text{l}_k,t}\right), \quad \forall \text{l}_k \in EL \tag{5.58}$$

$$\left|P^{\text{line}}_{t,e,\text{l}_k}\right| \leq P^{\text{line,max}}_{\text{l}_k} D_{\text{l}_k,t}, \quad \forall \text{l}_k \in EL \tag{5.59}$$

$$D_t = \left[\begin{array}{c|ccccc} 1 & 0 & 1 & \ldots & 1 \\ 1 & 1 & 0 & \ddots & \ldots \\ \ldots & \ldots & \ddots & \ddots & 1 \\ 1 & 1 & \ldots & 1 & 0 \end{array}\right] \tag{5.60}$$

where $EL = \{\text{l}_k \to (m_k, n_k)\,|k = 1, 2, \ldots, N_{\text{EL}}\}$. It is assumed that the N-1 contingency of power system only occurs in the existing transmission lines. Equations (5.58)–(5.59) indicate the power flow constraints of existing transmission lines under pre- or post-contingency state. D_t (5.60) denotes a $N_{\text{EL}} \times (N_{\text{EL}} + 1)$ contingency matrix at planning stage t, which contains 0 and 1 as the state of lines (0 indicates the line on outage while 1 indicates in service). The first column of D_t represents the pre-contingency condition, while the others represent the N-1 contingency states, in which one line is on outage. If line l_k during planning stage t is on outage, $D_{\text{l}_k,t} = 0$, Equation (5.58) will be relaxed while (5.59) setting the flow to zeros.

5.3.1.4 Compact Form of the MCC Model

The MCC model above can be presented as a MILP problem in a compact form as follows:

$$\min_{\mathbf{x},\mathbf{z}^k} C = \mathbf{cx} + \mathbf{dz}^0 \tag{5.61}$$

$$\text{s.t.} \quad \mathbf{ax} \leq \mathbf{b} \tag{5.62}$$

$$\mathbf{ex} + \mathbf{fz}^k \leq \mathbf{g}, \quad \forall k = 0, 1, 2, \ldots, N_{\text{EL}} \tag{5.63}$$

$$\mathbf{u}^k\mathbf{x} + \mathbf{v}^k\mathbf{z}^k \leq \mathbf{r}^k, \quad \forall k = 0, 1, 2, \ldots, N_{\text{EL}} \tag{5.64}$$

where $\mathbf{x}$, $\mathbf{z}^k$ are the decision variables. Specially, $\mathbf{x}$ includes the binary variables (corresponding to X, Y, and Z), as well as the integer variable (corresponding to U). $\mathbf{z}^k$ denotes the continuous variables (corresponding to P^{coal}, f^{supply} and P^{cur}) in all pre- and post-contingency states. For all line contingency states, $k = 0, 1, 2, \ldots, N_{\text{EL}}$. We point out that $k = 0$ denotes the pre-contingency state, i.e., normal operating condition. Equation (5.62) represents the constraints of sequential investment states of new assets. Equation (5.63) represents the operational constraints, including constraints of power system, natural gas fuel supply availability, and physical limits of interconnected media. Equation (5.64) corresponds to transmission N-1 contingency constraints, in which the coefficient matrices $\mathbf{u}^k$, $\mathbf{v}^k$, $\mathbf{r}^k$ change under the different contingency states.

5.3.2 *Solution Methodology*

For the large-scale system, to check all possible outage of the transmission lines will greatly increase the problem size. In this context, to improve the computational efficiency, an iterative Benders decomposition (IBD) method is presented to solve the proposed MCC problem (5.61)–(5.64), which is divided into master investment problem, pre-contingency check subproblem, N-1 check subproblem, and operation optimal subproblem. The flowchart of the proposed algorithm is shown in Fig. 5.12, which can be summarized as the following steps:

- Step 1: Data process: scenarios reduction method based on AP clustering (see Sect. 5.3.2.1).
- Step 2: The master investment problem finds the least cost investment plans over the planning horizon (see Sect. 5.3.2.2).
- Step 3: The feasibility cut will be generated in case of the violation in the pre-contingency check subproblem, which will be sent to the master investment problem to change the previous plan decision (see Sect. 5.3.2.3).
- Step 4: Then a contingency screening method based on line outage distribution factors (LODFs) is applied to estimate the post-contingency flow to obtain core contingency set C_C. If any contingency exists, the corresponding N-1 check subproblem under the most severe contingency will be conducted and send a reliability cut to master problem. Once all existing violations are removed, a reliable planning solution for electricity-gas systems is obtained, which satisfies all the N-1 contingency (see Sect. 5.3.2.4).
- Step 5: Finally, the operation optimal subproblem under all selected scenarios will be executed to find an upper bound. If the stopping criterion is satisfied, the final co-planning solution is obtained (see Sect. 5.3.2.5).

5.3.2.1 Scenario Reduction Method Based on AP Clustering

Though scenario-based method is widely used in stochastic optimization to capture the fluctuation of uncertain variables, large-scale scenarios will increase the computation burden. Therefore, it is essential to apply clustering technique to obtain a set of "clustered" and representative scenarios. The k-means algorithm is very simple and widely used in solving many practical problems (Wang et al. 2018; Grigoras and Scarlatache 2015). However, it is very sensitive to the initial partitions and extremely relies on the number of clusters. On the contrast, affinity propagation (AP) enables to

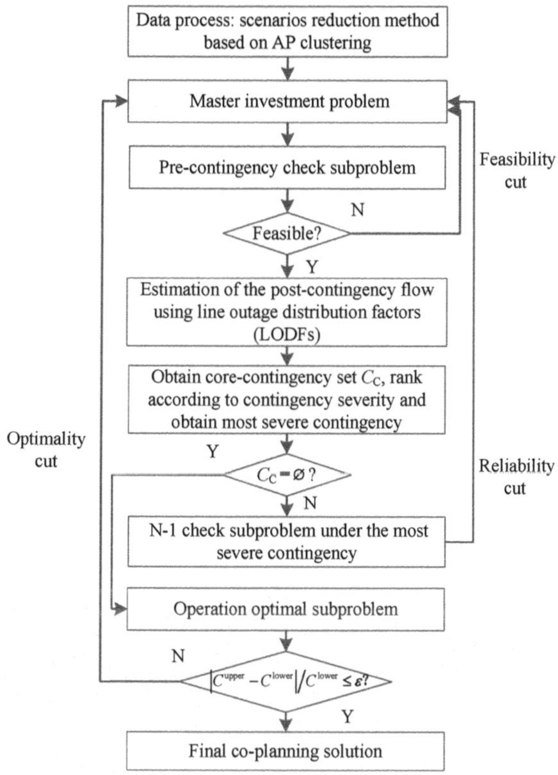

Fig. 5.12 Flowchart of the proposed algorithm

avoid most of the poor solutions caused by unlucky initializations, since it simultaneously regards all data points as candidate centers and gradually identifies clusters (Frey and Dueck 2007). Although AP still does not guarantee global optimum, the authors of Frey and Dueck (2007) have demonstrated its consistent superiority over most of clustering techniques by several experiments.

The AP clustering can be summarized by the following procedure.

- Step 1: Initialization of algorithm, set $r(i,k) = a(i,k) = 0$. where responsibility matrix $r(i,k)$ indicates how strongly each data point i favors the candidate exemplar k over other candidate exemplars k'; availabilities matrix $a(i,k)$ indicates to what degree each candidate exemplar k is available as a cluster center for the data point i.
- Step 2: Update responsibility matrix $r(i,k)$.

$$r_{t+1}(i,k) = s(i,k) - \max_{k \neq k'} \left\{ a_t\left(i,k'\right) + s\left(i,k'\right) \right\}$$

where similarity matrix $s(i, k)$ indicates how well the data point k is suited to be the exemplar for data point i, which is usually set to a negative squared error (Euclidean distance).

- Step 3: Update availability matrix $a(i, k)$.

$$a_{t+1}(i, k) = \min\left\{0, r_{t+1}(k, k) + \sum_{i' \notin \{i,k\}} \max\left\{0, r_{t+1}\left(i', k\right)\right\}\right\}$$

$$a_{t+1}(k, k) = \sum_{i' \neq k} \max\left\{0, r_{t+1}\left(i', k\right)\right\}$$

- Step 4: Introduce a damping factor λ to avoid oscillation.

$$r_{t+1}(i, k) \leftarrow (1 - \lambda)\, r_{t+1}(i, k) + \lambda r_t(i, k)$$

$$a_{t+1}(i, k) \leftarrow (1 - \lambda)\, a_{t+1}(i, k) + \lambda a_t(i, k)$$

- Step 5: Repeat Steps 2–4 until the matrices $r(i, k)$ and $a(i, k)$ are stable or the maximum number of iteration is reached.

In this study, the historical data includes 8760 hourly total electricity loads, total non-power gas demands, as well as each wind energy source located in each wind farm. After applying AP clustering, a set of representative scenarios will be obtained.

5.3.2.2 Master Investment Problem

The master investment problem (5.65) optimizes the planning decision over the planning horizon.

$$\min_{\mathbf{x}, \varphi} C^{\text{lower}} = \mathbf{cx} + \varphi \tag{5.65}$$

$$\text{s.t. } \mathbf{ax} \leq \mathbf{b} \tag{5.66}$$

$$\varphi > 0 \tag{5.67}$$

where φ is a piecewise function of the optimal value of the operation optimal subproblem. C^{lower} is a lower bound of the whole problem and will be updated iteratively by the operation optimal subproblem.

5.3.2.3 Pre-contingency Check Subproblem

Once the initial planning decision is identified by the above master investment problem, the pre-contingency check subproblem will be conducted. The objective (5.68) is to minimize the load curtailment for balancing purposes in the case of increasing load and wind power at each planning stage. The pre-contingency check subproblem is subjected to the operational constraints and pre-contingency constraints of existing transmission lines. However, to check all possible realization of uncertain load and wind power at each planning stage will greatly increase the computation complexity. To reduce the computational effort, the scenarios with the minimum wind energy or maximum load will be selected (Ugranli and Karatepe 2016; Kamyab et al. 2014).

$$\min_{\mathbf{z}^0, \mathbf{s}^0} h^0 = \mathbf{1}^T \cdot \mathbf{s}^0 \tag{5.68}$$

$$\text{s.t. } \mathbf{x} = \mathbf{x}^* \longrightarrow (\gamma) \tag{5.69}$$

$$\mathbf{ex} + \mathbf{fz}^0 \leq \mathbf{g} + \mathbf{s}^0 \tag{5.70}$$

$$\mathbf{u}^0\mathbf{x} + \mathbf{v}^0\mathbf{z}^0 \leq \mathbf{r}^0 \tag{5.71}$$

where $\mathbf{1}^T$ is the vector of ones; $\mathbf{s}^0$ is the slack variable corresponding to the power balance constraints and nodal balance equation of natural gas network; $h^0 > 0$ means that violations occur in the subproblem. $\mathbf{x}^*$ is the solution of initial plan obtained from the master investment problem; γ is the dual variable; $\mathbf{u}^0$, $\mathbf{v}^0$, $\mathbf{r}^0$ are the coefficient matrices of the transmission N-1 constraints under the normal operating condition.

$$h^0 + \gamma(\mathbf{x} - \mathbf{x}^*) \leq 0 \tag{5.72}$$

In order to eliminate the violations, the feasibility cut (5.72) will be generated and sent to the master investment problem to update the initial solution to ensure no load curtailment.

5.3.2.4 N-1 Check Subproblem

For the large-scale system, to check all possible outage of the transmission lines will greatly increase the computational burden. In fact, only a small part of the N-1 accidents are actually binding contingencies. If these core contingencies can be identified in advance, the amount of redundant calculations can be greatly reduced. For this reason, we utilize a contingency screening method based on LODFs to quickly calculate the post-contingency flow of lines and create the core contingency set. LODFs are defined as the impact of a line outage on post-contingency flows of other lines. Although there are several different formulations to calculate LODFs (Guler et al. 2007; Guo et al. 2009; Wood and Wollenberg 2012), they are all derived from the linear DC power flow equation. In this section, we use the direct calculation method of LODFs (Guo et al. 2009) to calculate post-contingency flows by the following equations:

$$PTDF_{\mathrm{l}_g,\mathrm{l}_k} = B_{\mathrm{l}_g,\mathrm{l}_g}\Psi_{\mathrm{l}_g}^T\left[Y'\right]^{-1}\Psi_{\mathrm{l}_k} \tag{5.73}$$

$$PTDF_{\mathrm{l}_k,\mathrm{l}_k} = B_{\mathrm{l}_k,\mathrm{l}_k} \Psi_{\mathrm{l}_k}^{T} \left[Y'\right]^{-1} \Psi_{\mathrm{l}_k} \tag{5.74}$$

$$LODF_{\mathrm{l}_g,\mathrm{l}_k} = PTDF_{\mathrm{l}_g,\mathrm{l}_k} \left(1 - PTDF_{\mathrm{l}_k,\mathrm{l}_k}\right)^{-1} \tag{5.75}$$

$$P_{\mathrm{l}_g,\mathrm{l}_k}^{\mathrm{line}} = P_{\mathrm{l}_g}^{\mathrm{line}} + LODF_{\mathrm{l}_g,\mathrm{l}_k} P_{\mathrm{l}_k}^{\mathrm{line}} \tag{5.76}$$

where $PTDF_{\mathrm{l}_g,\mathrm{l}_k}$ is Power Transfer Distribution Factors, defined as the impact of injecting 1 MW at bus m_k and withdrawing 1 MW from bus n_k on flow in line l_g. $LODF_{\mathrm{l}_g,\mathrm{l}_k}$ will be equal to -1 when $\mathrm{l}_k = \mathrm{l}_g$, since the line flow is set to 0 as it is on outage. We can utilize (5.76) calculate the post-contingency flow of line l_g, i.e., $P_{\mathrm{l}_g,\mathrm{l}_k}^{\mathrm{line}}$ as line l_k is on outage.

To obtain the core contingency set, we have

$$\Gamma_{\mathrm{l}_g,\mathrm{l}_k} = (P_{\mathrm{l}_g,\mathrm{l}_k}^{\mathrm{line}} - P_{\mathrm{l}_g}^{\mathrm{line,max}})/P_{\mathrm{l}_g}^{\mathrm{line,max}}, \quad \forall \mathrm{l}_k \in EL \tag{5.77}$$

$$\Delta_{\mathrm{l}_k} = \left\{\mathrm{l}_g \middle| \Gamma_{\mathrm{l}_g,\mathrm{l}_k} > 0\right\}, \quad \forall \mathrm{l}_k \in EL \tag{5.78}$$

$$\mathrm{l}_k = \begin{cases} \sum\limits_{\mathrm{l}_g \in \Delta \mathrm{l}_k} \Gamma_{\mathrm{l}_g,\mathrm{l}_k} / |\Delta \mathrm{l}_k|, & \Delta_{\mathrm{l}_k} \neq \varnothing, \ \forall \mathrm{l}_k \in EL \\ 0, & \Delta_{\mathrm{l}_k} = \varnothing, \ \forall \mathrm{l}_k \in EL \end{cases} \tag{5.79}$$

$$C_{\mathrm{C}} = \left\{\mathrm{l}_k \middle| \, \mathrm{l}_k > 0\right\} \tag{5.80}$$

Equations (5.77)–(5.78) select overload lines when an outage l_k occurs. Equation (5.79) evaluates the severity under each contingency, where a higher Λ represents a more severe contingency which should be taken into consideration first. The core contingency set C_{C} can be obtained via (5.80). Once the most severe contingency state $k = k'$ in the core contingency set is identified, N-1 check subproblem (5.81) will be carried out to obtain the corresponding objective value and dual variable.

$$\min_{\mathbf{z}^{k'},\mathbf{s}^{k'}} h^{k'} = \mathbf{1}^T \cdot \mathbf{s}^{k'} \tag{5.81}$$

$$\text{s.t. } \mathbf{x} = \mathbf{x}^* \longrightarrow (\boldsymbol{\pi}) \tag{5.82}$$

$$\mathbf{ex} + \mathbf{fz}^{k'} \leq \mathbf{g} + \mathbf{s}^{k'} \tag{5.83}$$

$$\mathbf{u}^{k'}\mathbf{x} + \mathbf{v}^{k'}\mathbf{z}^{k'} \leq \mathbf{r}^{k'} \tag{5.84}$$

where $\mathbf{s}^{k'}$ is the slack variable; $\boldsymbol{\pi}$ is the dual variable; $\mathbf{u}^{k'}$, $\mathbf{v}^{k'}$, $\mathbf{r}^{k'}$ are the coefficient matrices of the transmission N-1 constraints under the most severe contingency state k'.

According to the standard N-1 criterion, the power system should be planned and operated in a way to be able to supply all loads in case of one single outage in system components. In our work, we assume that no gas demands is allowed to be cut off during the N-1 contingency in power system. Therefore, the reliability cut (5.85) which is similar to the feasibility cut (5.72), is generated and sent to master investment problem to update the plan decision.

$$h^{k'} + \boldsymbol{\pi}(\mathbf{x} - \mathbf{x}^{*}) \leq 0 \tag{5.85}$$

5.3.2.5 Operation Optimal Subproblem

Once the reliable planning solution is obtained, the operation optimal subproblems (5.86) under the normal operating condition for all selected scenarios at each planning stage will be carried out, which are subjected to the operational constraints and pre-contingency constraints.

$$\min_{\mathbf{z}^0} \; q = \mathbf{d}\mathbf{z}^0 \tag{5.86}$$

$$\text{s.t.} \;\; \mathbf{x} = \mathbf{x}^{*} \longrightarrow (\boldsymbol{\beta}) \tag{5.87}$$

$$\mathbf{e}\mathbf{x} + \mathbf{f}\mathbf{z}^0 \leq \mathbf{g} \tag{5.88}$$

$$\mathbf{u}^0\mathbf{x} + \mathbf{v}^0\mathbf{z}^0 \leq \mathbf{r}^0 \tag{5.89}$$

where $\boldsymbol{\beta}$ is the dual variable; the optimal value q represents the sum of operation cost and curtailed wind energy cost under the normal operating condition, which will be added into the investment cost to give the total co-planning cost $C^{\text{upper}} = \mathbf{c}\mathbf{x}^{*} + q$. If the solution is not optimal, the optimality cut (5.90) will be generated and sent to master investment problem.

$$q + \boldsymbol{\beta}(\mathbf{x} - \mathbf{x}^{*}) \leq \varphi \tag{5.90}$$

Through the above iterative process to solve the proposed MCC model, an economic, reliable, and fuel supply feasible solution for long-term co-planning of EGS will be obtained.

5.3.3 *Case Studies*

To illustrate the performance, the proposed MCC model is tested on a modified Garver six-bus power system interconnected with a seven-node gas system, and a modified IEEE 118-bus system interconnected with a 14-node gas system. The simulation computation is completed in MATLAB using CPLEX solver.

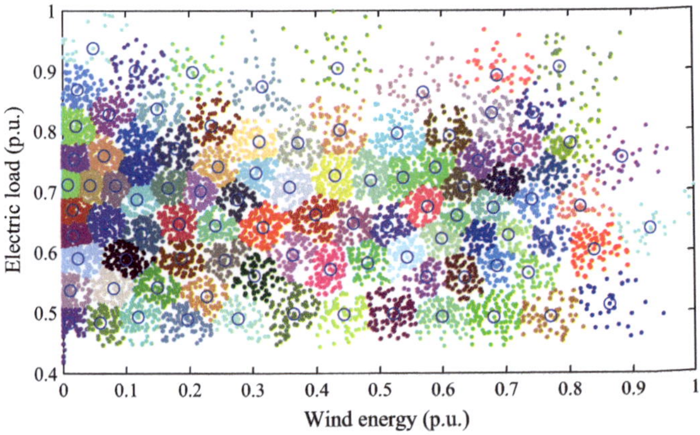

Fig. 5.13 Reduced data set using AP clustering

Table 5.15 Comparison of clustering performance of AP and k-means

Item	CalinskiHarabasz (CH)	DaviesBouldin (DB)	Silhouette(Sil)
AP	1.1485×10^4	0.7610	0.5539
k-means	1.1099×10^4	0.8163	0.5106

The historical data of 8760 hourly loads and wind energy is obtained from AEMO (Australian Energy Market Operator 2019). After using AP clustering, the original scenarios are classified into 96 groups, as shown in Fig. 5.13, in which the circle labels represent the chosen scenarios. To visually demonstrate the effect of AP clustering, Fig. 5.13 contains only two-dimensional variables. Different from k-means, AP clustering selects a sample from each group as the cluster center.

To evaluate the performance of AP clustering, we use the following three popular evaluation metrics by invoking the *evalclusters* function in MATLAB.

The comparison of clustering performance of AP and k-means is shown in Table 5.15. Index CH equals inter-cluster distances divided by proximity of intra-cluster. Therefore, the larger CH, the closer the elements intra-cluster, and more dispersed inter-cluster. Index DB calculates the similarity of elements inter-cluster, so a lower DB corresponds to the better clustering results. Index Sil measures the similarity of a element to its cluster over other clusters. And a higher Sil indicates the elements favor their current clusters more strongly. In this context, AP performs better than k-means.

5.3.3.1 The Modified Garver 6-Bus System and a 7-Node Gas Network

The modified Garver 6-bus system interconnected with a 7-node natural gas network is depicted in Fig. 5.14. The modified Garver 6-bus system includes 1 coal unit, 2

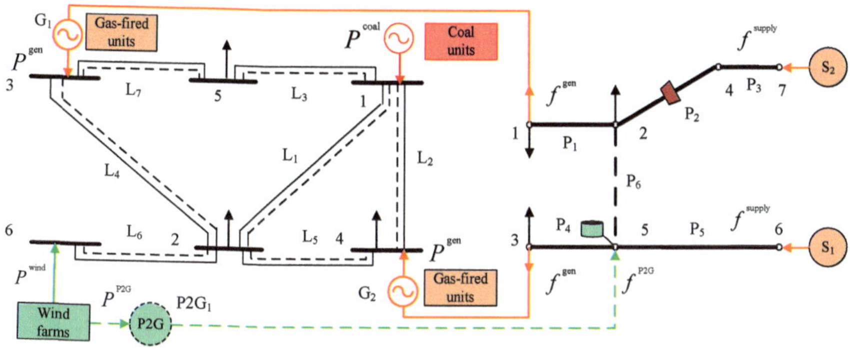

Fig. 5.14 A six-bus power system interconnected with a seven-node gas system

Table 5.16 Data of electricity lines

No.	From bus	To bus	Reactance (p.u.)	Capacity (MW)	Capital cost (M$/MW)
L_1	1	2	0.4	100	4
L_2	1	4	0.6	100	6
L_3	1	5	0.2	100	2
L_4	2	3	0.2	100	2
L_5	2	4	0.4	100	4
L_6	2	6	0.3	100	3
L_7	3	5	0.2	100	2

gas-fired units, 1 wind farm, 7 lines, and 3 electricity loads. In this study, we only consider the reinforcement of lines, while the construction of new lines between two buses where there was no original lines is not included (Ugranli and Karatepe 2016; Zhang et al. 2012). In the modified 7-node natural gas network, two gas sources S_1–S_2, respectively, feed the gas-fired units and non-power gas loads. The electricity-gas systems are connected by gas-fired units and candidate P2G plants. In general, the candidate locations for P2G plants are near the wind farms and connect to the nearest natural gas node. In this study, we assume that gas node 5 is closest to the wind farm. The imaginary lines in Fig. 5.14 indicate the candidate equipment. A set of 7 candidate transmission lines L_1–L_7, 1 candidate gas pipeline P_6(2–5), 2 candidate gas-fired units G_1–G_2 and 1 candidate P2G plants P2G_1(bus 6) are considered.

The data of electricity lines, natural gas pipelines, and generator units are listed in Tables 5.16, 5.17, and 5.18 (Garver 1970; Liu et al. 2009). The capital cost of candidate equipments are obtained from Zhang et al. (2015), He et al. (2018).

The existing gas-fired units include a 150-MW at bus 3 and a 200-MW at bus 4. The capacity of candidate gas-fired units is the same as the existing ones. The existing coal units include a 360-MW at bus 1. The efficiency of P2G plants is assumed to be 64% and the investment cost of P2G plants is 1.5 M$/MW Sterner (2009). This

Table 5.17 Data of natural gas pipelines

No.	From node	To node	Capacity (kcf)	Capital cost (M$)
P_1	1	2	8000	–
P_2	2	4	8000	–
P_3	4	7	10000	–
P_4	3	5	8000	–
P_5	5	6	10000	–
P_6	2	5	8000	100

Table 5.18 Data of generator units

No.	Bus	Type	Capacity (MW)	Capital cost (M$/MW)	Operational cost ($/MWh)
G_1	3	Gas-fired units	150	0.6	–
G_2	4	Gas-fired units	200	0.6	–
G_3	1	Coal units	360	–	23.8

investment covers the electrolyzer, the methanation, the gas compression, power electronics, piping, civil works, and control systems (Tichler et al. 2014).

A 20-year planning horizon is divided into five stages. The average growth rate of total electrical load, non-power gas demand and wind energy at each planning stage are assumed to be 10%, 10%, and 15%, respectively. The electricity load distribution factors are 0.4, 0.3, and 0.3 for loads at bus 2, bus 4, and bus 5, respectively, while the non-power gas distribution factors for nodes 1–3 are 0.25, 0.25, and 0.5. The discount rate is 8%. The lifespan is assumed to be 40 years for all candidate equipment. The gas fuel cost is 6.24 $/kcf Zhang et al. (2015).

(a) Effect of considering N-1 contingency

In this part, two cases are designed as the following and their planning results are presented in Table 5.19.

Case 1: Without considering N-1 contingency.

Case 2: Base case MCC problem considering N-1 contingency. A complete N-1 analysis is performed on each transmission line.

The symbol $G_{1,4}$ denotes the gas-fired units G_1 are constructed and installed at stage 4. As seen from Table 5.19, without considering N-1 contingency, only two gas-fired units, two transmission lines, as well as a 81-MW P2G plants are installed in Case 1. While in Case 2, two extra transmission line L_1(1–2), L_7(3–5) and gas pipeline P_6(2–5) are built when considering N-1 criterion for all transmission lines. In Case 2, once transmission line L_2(1–4) or L_5(2–4) is on outage, the electricity load at bus 4 is mainly met by gas-fired units G_2, which are supplied by gas supplier S_1. At stage 5, the gas supplier S_1 face a fuel supply shortage. Therefore, gas pipeline P_6(2–5) has to be installed to link up two-area natural gas system to compensate fuel supply shortage by delivering natural gas from gas supplier S_2.

Table 5.19 Optimal result with and without N-1 contingency

Constructed components	Case 1	Case 2
Gas-fired units	$G_{1,4}$, $G_{2,5}$	$G_{1,1}$, $G_{2,3}$
Transmission lines	$L_{3,1}$, $L_{4,3}$	$L_{1,5}$, $L_{3,1}$, $L_{4,1}$, $L_{7,3}$
Gas pipelines	–	$P_{6,5}$
P2G plants (MW)	$P2G_1(81)$	$P2G_1(81)$
Investment cost (M$)	282.3	573.6
Operation cost (M$)	4288.6	4283.7
Curtailed wind energy cost (M$)	2.3	2.6
Total co-planning cost (M$)	4572.1	4859.9

Table 5.20 Comparison of constructed components, total co-planning costs and wind power utilization

Constructed components	Model 1	Model 2
Gas-fired units	$G_{1,1}$, $G_{2,3}$	$G_{1,1}$, $G_{2,3}$
Transmission lines	$L_{1,5}$, $L_{3,1}$, $L_{4,1}$, $L_{7,3}$	$L_{1,5}$, $L_{3,1}$, $L_{4,1}$, $L_{6,4}$, $L_{7,3}$
Gas pipelines	$P_{6,5}$	$P_{6,5}$
P2G plants (MW)	$P2G_1(81)$	–
Investment cost (M$)	573.6	625.0
Operation cost (M$)	4283.7	4265.6
Curtailed wind energy cost (M$)	2.6	26.1
Total co-planning cost (M$)	4859.9	4916.7
Wind power utilization	97.25%	88.78%

In addition, the constructed components of Case 2 are almost built ahead of that of Case 1, to enable power system to satisfy N-1 criterion. Operation cost and curtailed wind energy cost of these two cases are near. Though the total co-planning cost of Case 1 outperforms that of Case 2, the optimal plan of Case 1 can not ensure power system safe and reliable.

(b) Effect of P2G plants

To evaluate the economic merits of P2G plants and its impacts on the co-planning, two following models are carried out.

Model 1: Base case MCC model.

Model 2: Without considering P2G plants.

Table 5.20 lists the comparison of the constructed components, total co-planning costs and wind power utilization. These two models obtain some same constructed components, including two gas-fired units, four transmission lines, and one gas pipeline. The main difference of optimum results reflect in the installation of P2G plants and transmission lines L_6(2–6). In Model 1, due to the physical limit of existing transmission line L_6, 81-MW P2G plants (consisting of 2-MW, 17-MW, 15-MW,

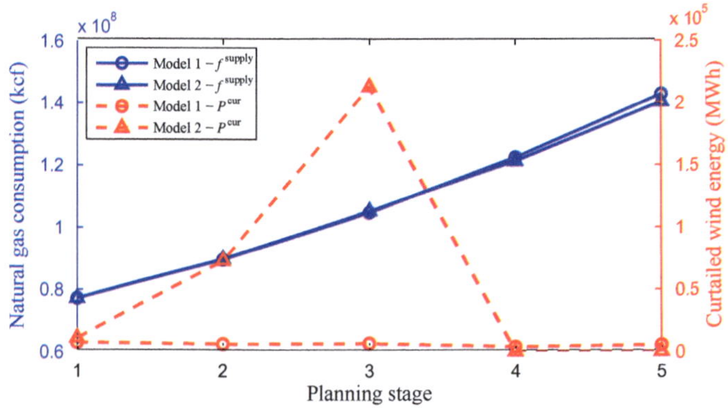

Fig. 5.15 Comparison of natural gas consumption and curtailed wind energy at each planning stage

23-MW and 24-MW installed at stage 1 to stage 5, respectively) are built to economically convert excessive wind energy into synthetic natural gas (SNG). While in Model 2, the relatively expensive candidate transmission line L_6 is installed at stage 4 to utilize growing wind energy and prevent wind spillage. It should be mentioned that the candidate L_6 is not installed at stage 1, since during that time the penalty cost for curtailed wind energy is lower than investment cost of installing a new line. For this reason, Model 1 presents an obvious advantage in the investment cost, compared with Model 2.

In addition, Model 1 pays less 23.5 M$ penalty cost for curtailed wind energy than Model 2. As shown in Fig. 5.15, due to the physical limit of installed P2G plants, a small fraction of wind energy will be curtailed in those scenarios where the wind energy is high but the probability of occurrence is low. While in Model 2, the surplus wind energy has to be curtailed until the candidate L_6 is constructed at stage 4.

Note that from stage 1 to stage 3, the natural gas consumption of Model 1 is slightly lower than that of Model 2, with the help of P2G plants to produce SNG. However, the situation is opposite from stage 4 to stage 5. This is because, during those stages, the wind energy in Model 2 is completely utilized to meet the increasing demands. While in Model 1, the energy loss during the P2G process causes more energy consumption of generation units and gas suppliers, thus leading to a little higher operation cost. Nevertheless, the direct wind power utilization during the planning horizon is increased from 88.78 to 97.25% due to the installation of P2G plants.

To make a summary, this example illustrates that the proposed MCC model considering P2G plants can significantly improve wind power utilization by nearly 8% and lead to a lower total co-planning cost.

(c) Effect of unit penalty cost of curtailed wind energy

Figure 5.16a, b explores the effect of unit penalty cost of curtailed wind energy on the optimal results. As the unit penalty cost increases, larger capacity of P2G plants are

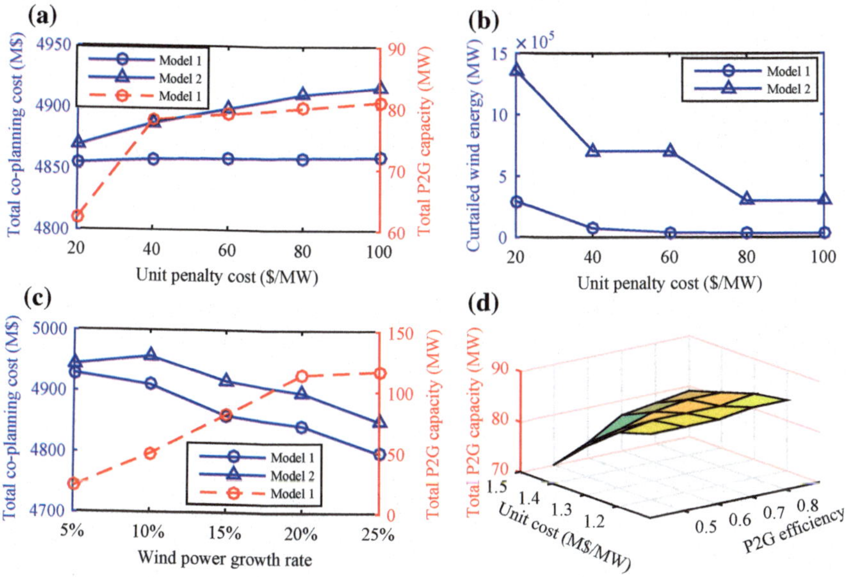

Fig. 5.16 Comparison of total co-planing cost, P2G capacity and curtailed wind energy under different parameters

selected to utilize the surplus wind energy to avoid higher penalty cost for curtailing wind energy in Model 1. Therefore, the total co-planning cost of Model 1 shows a rising trend while curtailed wind energy with a downward trend. While in Model 2, as the unit penalty cost increases, the candidate transmission line L_6 has to be constructed in advance from stage 4 to stage 2.

Therefore, with P2G plants consideration, the total co-planning cost and curtailed wind energy of Model 1 are always lower than those of Model 2.

(d) Effect of wind growth rate

Figure 5.16c illustrates the total co-planning cost and total P2G capacity as a function of wind power growth rate. The total co-planning cost in Model 1 decreases when a higher growth rate of wind power is considered. This is because larger capacity of P2G plants will be built to utilize the wind energy to greatly reduce the operation cost. The total cost of Model 2 also shows a downward trend as a whole, but it is still higher than that of Model 1.

(e) Effect of P2G efficiency and unit cost

To explore the influence of P2G efficiency and unit cost of P2G plants on the optimum results, the simulations under different efficiency and unit cost are carried out. With a higher P2G efficiency, less energy loss during the process of P2G will be caused. Therefore, a higher capacity of P2G plants will help to reduce the curtailed wind power energy, as shown in Fig. 5.16d. In addition, with the advancement of P2G

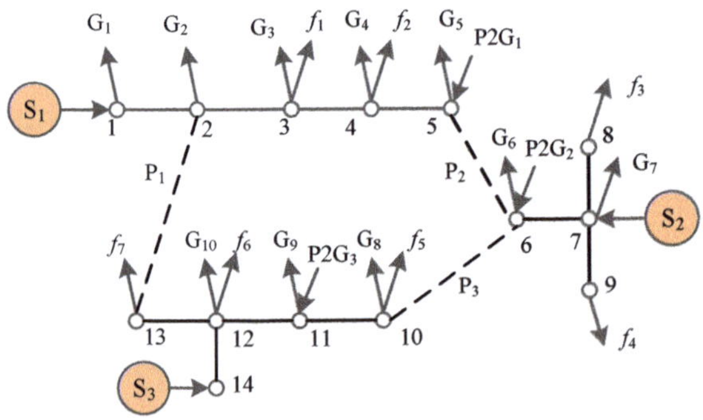

Fig. 5.17 Three-area 14-node gas network

Table 5.21 Candidate transmission lines of IEEE 118-bus

No.	Lines	No.	Lines	No.	Lines	No.	Lines	No.	Lines	No.	Lines	No.	Lines
L_1	3–5	L_5	30-17	L_9	30–38	L_{13}	82–83	L_{17}	100–103	L_{21}	71–73	L_{25}	90–91
L_2	8–9	L_6	8–30	L_{10}	55–56	L_{14}	83–85	L_{18}	17–113	L_{22}	86–87	L_{26}	2–12
L_3	8-5	L_7	26–30	L_{11}	38–65	L_{15}	85–86	L_{19}	12–117	L_{23}	110–111	–	–
L_4	11–12	L_8	29–31	L_{12}	77–78	L_{16}	91–92	L_{20}	9–10	L_{24}	38-37	–	–

technology, a lower unit cost will be beneficial to the installation of P2G plants, which will be more economical and competitive compared with the transmission line.

5.3.3.2 Modified IEEE 118-Bus System and 14-Node Gas Network

A modified IEEE 118-bus system interconnected with 14-node gas network is applied here to further demonstrate the applicability of the proposed MCC model on large systems, as shown in Fig. 5.17. The modified IEEE 118-bus system includes 42 coal units, 10 gas-fired units, 3 wind farms, and 186 lines. Note that in the previous case study, every corridor where a transmission line exists is considered to install another candidate one. However, in a large-scale system, based on either previous operational experience or physical and regulatory limits, it is practical to form a candidate line set in which only a few valid or important candidate lines are listed (Wood and Wollenberg 2012; Zhang et al. 2012). In this case study, 26 transmission lines are selected as candidate ones, as shown in Table 5.21.

Three-area 14-node gas networks feed 10 candidate gas-fired units G_1–G_{10} and non-power gas loads $f_1 - f_7$, and are connected with 3 candidate P2G plants $P2G_1$–$P2G_3$. Three candidate gas pipelines (imaginary lines P_1–P_3) are considered to link the delivery areas. The data of candidate gas pipelines, gas-fired units and P2G plants

Table 5.22 Candidate gas pipelines

No.	From node	To node	Capacity (kcf)	Capital cost (M$)
P_1	2	13	8000	90
P_2	5	6	8000	90
P_3	6	10	10000	100

Table 5.23 Candidate gas-fired units

No.	Node	Bus	Capacity (MW)	Capital cost (M$/MW)
G_1	1	6	100	0.6
G_2	2	12	200	0.65
G_3	3	18	200	0.65
G_4	4	25	100	0.6
G_5	5	32	100	0.6
G_6	6	56	200	0.65
G_7	7	62	100	0.6
G_8	10	89	200	0.65
G_9	11	100	200	0.65
G_{10}	12	113	100	0.6

Table 5.24 Candidate P2G plants

No.	Node	Bus	Efficiency	Capital cost (M$/MW)
$P2G_1$	5	73	0.64	1.5
$P2G_2$	6	91	0.64	1.5
$P2G_3$	11	117	0.64	1.5

are listed in Tables 5.22, 5.23 and 5.24. Other network parameters can be found in Zhang et al. (2015), He et al. (2018).

In this study, a set of 26 candidate transmission lines L_1–L_{26}, 3 candidate gas pipeline P_1–P_3, 10 candidate gas-fired units G_1–G_{10} and 3 candidate P2G plants $P2G_1$–$P2G_3$ are considered. Three cases are presented to illustrate the availability and effectiveness of the proposed model.

Case 1: Without considering N-1 contingency.

Case 2: Base case MCC problem considering N-1 contingency using LODFs.

Case 3: MCC problem with all contingencies.

The comparison of constructed components, total co-planning cost and computation time in Case 1–3 are summarized in Table 5.25. In Case 1, two new gas-fired units and six new transmission lines are constructed to meet the increasing electricity loads. New gas pipeline P_3 is selected to deliver natural gas from gas supplier S_2 to compensate the fuel supply shortages in S_3. Moreover, 102-MW P2G plants are

Table 5.25 Optimal result of constructed components, co-planning cost, and computation time

Constructed components	Case 1	Case 2	Case 3
Gas-fired units	$G_{5,5}$, $G_{6,5}$	$G_{1,2}$, $G_{2,5}$, $G_{10,5}$	$G_{1,2}$, $G_{2,5}$, $G_{10,5}$
Transmission lines	$L_{3,2}$, $L_{9,2}$, $L_{11,4}$,	$L_{2,4}$, $L_{3,2}$, $L_{4,4}$, $L_{9,2}$, $L_{11,4}$,	$L_{2,4}$, $L_{3,2}$, $L_{4,4}$, $L_{9,2}$, $L_{11,4}$,
	$L_{20,5}$, $L_{21,4}$, $L_{25,4}$	$L_{20,4}$, $L_{21,2}$, $L_{24,5}$, $L_{25,3}$, $L_{26,3}$	$L_{20,4}$, $L_{21,2}$, $L_{24,5}$, $L_{25,3}$, $L_{26,3}$
Gas pipelines	$P_{3,3}$	$P_{3,3}$	$P_{3,3}$
P2G plants (MW)	$P2G_1$(102)	$P2G_1$(102)	$P2G_1$(102)
Investment cost (M$)	320.6	382.6	382.6
Operation cost (M$)	21479.5	21470.0	21470.0
Curtailed wind energy cost (M$)	26.3	25.7	25.7
Total co-planning cost (M$)	21800.1	21852.6	21852.6
Computation time (sec)	311.9	852.1	18501.3

Table 5.26 Selected lines for contingency analysis using LODFs

Item	Selected lines for contingency analysis
C_C	4–5, 8–9, 8-5, 9–10, 2–12, 15–17, 23–24, 25–27, 26–30, 38–37

installed in the first wind farms to utilize the surplus wind energy. Compare with Case 1, extra one gas-fired units and four new transmission lines are added in Case 2. For this reason, more investment cost is spent in Case 2 for enabling power system to satisfy the N-1 criterion.

In Case 3, the MCC problem with all contingencies is also solved to compare its results and computation time with MCC problem using LODFs to obtain core contingency set (Case 2). Final results show that these two methods select the same optimal plans with the same total co-planning costs. However, Case 2 is more than 20 times faster than Case 3. This is because only 10 contingency lines are selected after using LODFs for contingency analysis, as shown in Table 5.26, which are reduced by over 94%.

The number of iteration in Case 2 is 43 with a relative rap of $\varepsilon = 0.5\%$. The convergence performance of the decomposition method is acceptable, as shown in Fig. 5.18.

Therefore, case studies above verify the availability of the proposed MCC model and the effectiveness of the proposed IBD method. The contingency handling strategy possesses high computational efficiency.

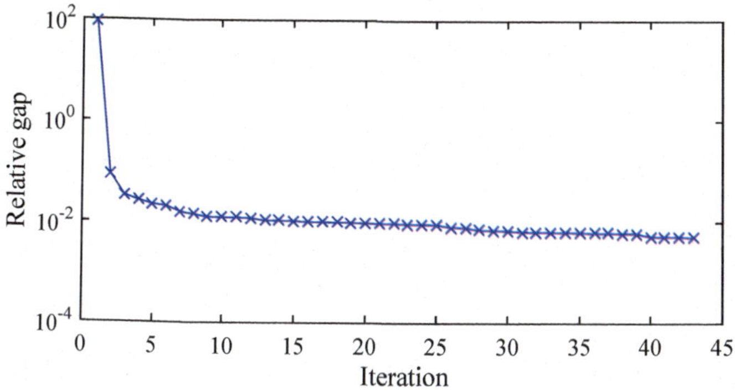

Fig. 5.18 Iteration convergence

5.4 Summary

This chapter has presented the study of multi-stage contingency-constrained co-planning (MCC) for electricity-gas systems (EGS) interconnected with gas-fired units and P2G plants using iterative Benders decomposition (IBD). The simulation results carried out on two different scales of EGS have demonstrated the following conclusions:

First, the MCC model developed for EGS can be tackled by the proposed IBD method, considering the long-term co-planning with the short-term operation constraints. The simulation results demonstrate that the MCC model considering P2G plants can significantly improve wind power utilization by nearly 8% and lead to a lower total co-planning cost, compared with the model without P2G plants. According to the simulation results of the sensitive analyses, the unit penalty cost of curtailed wind energy, wind growth rate, and P2G efficiency have influences on final co-planning solution of EGS.

Moreover, the simulation results illustrate that affinity propagation clustering performs better than k-means in reducing the number of initial scenarios to an acceptable level without any pre-specification. The contingency accidents can be reduced by over 94% after using LODFs, which possessing high computational efficiency. Furthermore, the number of iteration in the large-scale test system is 43 with the computation time of 852.1 s, which is an accepted level for the long-term co-planning.

The proposed MCC model could be adopted by electricity-gas systems planners or regional regulatory authorities for long-term coordination of electricity-gas systems. In addition, the IBD method could be extended to tackle the N-*k* contingency-constrained problem in our future work.

References

Abbes D, Martinez A, Champenois G (2014) Life cycle cost, embodied energy and loss of power supply probability for the optimal design of hybrid power systems. Math Comput Simul 98:46–62

Ai X, Wen J, Wu T, Lee WJ (2013) A discrete point estimate method for probabilistic load flow based on the measured data of wind power. IEEE Trans Ind Appl 49:2244–2252

Australian Energy Market Operator. http://www.aemo.com.au/

Barati F et al (2015) Multi-period integrated framework of generation, transmission, and natural gas grid expansion planning for large-scale systems. IEEE Trans Power Syst 30(5):2527–2537

Benders JF (1962) Partitioning procedures for solving mixed-variables programming problems. Numer Math 4(1):238–252

Borraz-Sánchez C, Ríos-Mercado RZ (2009) Improving the operation of pipeline systems on cyclic structures by tabu search. Comput Chem Eng 33(1):58–64

Colbertaldo P, Guandalini G, Campanari S (2018) Modelling the integrated power and transport energy system: the role of power-to-gas and hydrogen in long-term scenarios for Italy. Energy 154:592–601

Dong RT, Xu JP (2015) Impact of differentiated local subsidy policies on the development of distributed energy system. Energy Build 101:45–53

Duran MA, Grossmann IE (1986) An outer-approximation algorithm for a class of mixed-integer nonlinear programs. Math Program 36(3):307–339

Frey BJ, Dueck D (2007) Clustering by passing messages between data points. Science 315(5814):972–976

Garver LL (1970) Transmission network estimation using linear programming. IEEE Trans Power Appar Syst 7:1688–1697

Grigoras G, Scarlatache F (2015) An assessment of the renewable energy potential using a clustering based data mining method. case study in Romania. Energy 81:416–429

Guandalini G, Campanari S, Romano MC (2015) Power-to-gas plants and gas turbines for improved wind energy dispatchability: energy and economic assessment. Appl Energy 147:117–130

Guler T, Gross G, Liu MH (2007) Generalized line outage distribution factors. IEEE Trans Power syst 22(2):879–881

Guo JC, Fu Y, Li ZY, Shahidehpour M (2009) Direct calculation of line outage distribution factors. IEEE Trans Power Syst 24(3):1633–1634

Guo CX, Zhan JP, Wu QH (2012) Dynamic economic emission dispatch based on group search optimizer with multiple producers. Electr Power Syst Res 86:8–16

Guo L, Liu WJ, Cai JJ, Hong BW, Wang CS (2013) A two-stage optimal planning and design method for combined cooling, heat and power microgrid system. Energy Convers Manag 74:433–445

He C, Wu L, Liu TQ, Bie ZH (2018) Robust co-optimization planning of interdependent electricity and natural gas systems with a joint N-1 and probabilistic reliability criterion. IEEE Trans Power Syst 33(2):2140–2154

Hemmati R, Saboori H, Siano P (2017) Coordinated short-term scheduling and long-term expansion planning in microgrids incorporating renewable energy resources and energy storage systems. Energy 134:699–708

Hu Y, Bie ZH, Ding T, Lin YL (2016) An NSGA-II based multi-objective optimization for combined gas and electricity network expansion planning. Appl Energy 167:280–293

Jiang C, Han X, Liu G, Liu G (2008) A nonlinear interval number programming method for uncertain optimization problems. Eur J Oper Res 188:1–13

Jiang XS, Jing ZX, Li YZ, Wu QH, Tang WH (2014) Modeling and operation optimization of an integrated energy based direct district water-heating system. Energy 64:375–388

Jing ZX, Jiang XS, Wu QH, Tang WH, Hua B (2014) Modelling and optimal operation of a small-scale integrated energy based district heating and cooling system. Energy 73:399–415

Kamyab GR, Fotuhi-Firuzabad M, Rashidinejad M (2014) A PSO based approach for multi-stage transmission expansion planning in electricity markets. Int J Electr Power Energy Syst 54:91–100

Kelley JE Jr (1960) The cutting-plane method for solving convex programs. J Soc Ind Appl Math 8(4):703–712

Li MS, Wu QH, Ji TY, Rao H (2014b) Stochastic multi-objective optimization for economic-emission dispatch with uncertain wind power and distributed loads. Electr Power Syst Res 116:367–373

Li YZ, Wu QH, Li MS, Zhan JP (2014a) Mean-variance model for power system economic dispatch with wind power integrated. Energy 72:510–520

Liu C, Shahidehpour M, Fu Y, Li ZY (2009) Security-constrained unit commitment with natural gas transmission constraints. IEEE Trans Power Syst 24(3):1523–1536

Lorenz E, Hurka J, Heinemann D, Beyer HG (2009) Irradiance forecasting for the power prediction of grid-connected photovoltaic systems. IEEE J Sel Top Appl Earth Obs Remote Sens 2:2–10

Majidi-Qadikolai M, Baldick R (2016) Integration of N-1 contingency analysis with systematic transmission capacity expansion planning: ERCOT case study. IEEE Trans Power Syst 31(3):2234–2245

Mancarella P (2014) MES (multi-energy systems): an overview of concepts and evaluation models. Energy 65(2):1–17

Mao Y (2007) Study on climate adaptability of human beings to thermal comfort in china. PhD thesis, Xian University of Architecture and Technology

Markowitz H (1952) Portfolio selection. J Financ 7:77–91

Moskalenko N, Lombardi P, Komarnicki P (2014) Multi-criteria optimization for determining installation locations for the power-to-gas technologies. In: Power and energy society general meeting. IEEE, pp 1–5

Odetayo B, MacCormack J, Rosehart WD, Zareipour H, Seifi AR (2018) Integrated planning of natural gas and electric power systems. Int J Electr Power Energy Syst 103:593–602

Parra D, Patel MK (2016) Techno-economic implications of the electrolyser technology and size for power-to-gas systems. Int J of Hydrogen Energy 41(6):3748–3761

Qiu J et al (2015) Multi-stage flexible expansion co-planning under uncertainties in a combined electricity and gas market. IEEE Trans Power Syst 30(4):2119–2129

Qiu J, Yang HM, Dong ZY et al (2016) A linear programming approach to expansion co-planning in gas and electricity markets. IEEE Trans Power Syst 31(5):3594–3606

Ramakumar R, Butler NG, Rodriguez AP, Venkata S (1993) Economic aspects of advanced energy technologies. Proc IEEE 81:318–332

Ren HB, Zhou WS, Gao WJ (2012) Optimal option of distributed energy systems for building complexes in different climate zones in China. Appl Energy 91:156–165

Roh JH, Shahidehpour M, Fu Y (2007) Market-based coordination of transmission and generation capacity planning. IEEE Trans Power Syst 22(4):1406–1419

Shao CC et al (2017) Integrated planning of electricity and natural gas transportation systems for enhancing the power grid resilience. IEEE Trans Power Syst 32(6):4418–4429

Sterner M (2009) Bioenergy and renewable power methane in integrated 100% renewable energy systems: limiting global warming by transforming energy systems. Kassel University Press GmbH

Tan ZF, Zhang HJ, Shi QS, Song YH, Ju LW (2014) Multi-objective operation optimization and evaluation of large-scale NG distributed energy system driven by gas-steam combined cycle in China. Energy Build 76:572–587

Thapar V, Agnihotri G, Sethi VK (2011) Critical analysis of methods for mathematical modelling of wind turbines. Renew Energy 36:3166–3177

Tichler R, Lehner M, Steinmller H, Koppe M (2014) Power-to-Gas: technology and Business Models. Springer Briefs in Energy

Ugranli F, Karatepe E (2016) Transmission expansion planning for wind turbine integrated power systems considering contingency. IEEE Trans Power Syst 31(2):1476–1485

Wang JX et al (2017) Review and prospect of integrated demand response in the multi-energy system. Appl Energy 202:772–782

U.S. Energy Information Administration, Monthly energy review. https://www.eia.gov/totalenergy/data/monthly/pdf/mer.pdf

Wang Y, Xia Q, Kang CQ (2011) Unit commitment with volatile node injections by using interval optimization. IEEE Trans Power Syst 26:1705–1713
Wang YJ, Dong ZY, Xu Y, Ma J, Zheng Y (2015) Enabling large-scale energy storage and renewable energy grid connectivity: a power-to-gas approach. CSEE J Power Energy Syst 35(14):3586–3595
Wang K, Qi XX, Liu HD, Song JK (2018) Deep belief network based k-means cluster approach for short-term wind power forecasting. Energy 165:840–852
Wood AJ, Wollenberg BF (2012) Power generation, operation, and control. Wiley
Wu QH, Lu Z, Li MS, Ji TY (2008) Optimal placement of facts devices by a group search optimizer with multiple producer. In: IEEE congress on evolutionary computation, CEC 2008 (IEEE World Congress on Computational Intelligence). IEEE, pp 1033–1039
Wu L, Shahidehpour M, Li ZY (2012) Comparison of scenario-based and interval optimization approaches to stochastic SCUC. IEEE Trans Power Syst 27:913–921
Zeng Q, Zhang BH, Fang JK, Chen Z (2017) A bi-level programming for multistage co-expansion planning of the integrated gas and electricity system. Appl Energy 200:192–203
Zhang H, Vittal V, Heydt GT, Quintero J (2012) A mixed-integer linear programming approach for multi-stage security-constrained transmission expansion planning. IEEE Trans Power Syst 27(2):1125–1133
Zhang XP, Shahidehpour M, Alabdulwahab AS, Abusorrah A (2015) Security-constrained co-optimization planning of electricity and natural gas transportation infrastructures. IEEE Trans Power Syst 30(6):2984–2993
Zheng JH, Wu QH, Jing ZX (2017) Coordinated scheduling strategy to optimize conflicting benefits for daily operation of integrated electricity and gas networks. Appl energy 192:370–381
Zhou M, Xia S, Li GY, Han X (2014) Interval optimization combined with point estimate method for stochastic security-constrained unit commitment. Int J Electr Power Energy Syst 63:276–284

Chapter 6
Optimal Operation of Large-Scale Integrated Energy Systems

Abstract The increasing share of variable renewable energy sources and the improving requirements on system security and reliability are calling for important changes in the LSIES. The synergies between energy supply networks are of great importance to satisfy the development of LSIES. Hence, this chapter presents the study of the coordinated scheduling strategy (CSS), in which, the models of the electricity network and gas network are developed in detail, and the operation constraints of the networks are fully considered. The purpose of the CSS is to optimize the conflicting benefits of the electricity network and gas network for daily operation of the LSIES, while satisfying the operation constraints. In the CSS, a multi-objective optimization algorithm is applied to obtain a Pareto-optimal solution set, and a multiple attribute decision analysis (MADA) using interval evidential reasoning (IER) is developed to determine a final optimal daily operation solution for the LSIES.

Keywords Optimal operation · Coordinated scheduling strategy · Conflict benefits · Energy trading game

6.1 Introduction to Operation of the LSIES

Until now, worldwide demand for natural gas has increased from 32% in 2007 to 39% in 2009 (Üster and Dilaveroğlu 2014), and it is estimated to grow at a rate of 2.9–3.2% per year until 2030 (EIA-US 2011). With the high integration of natural gas in electricity power systems, the synergies between the gas network and electricity network are dramatically increasing in order to reduce the operation cost and satisfy the stringent environmental regulations (Chaudry et al. 2014).

As a significant kind of gas consumers, the gas-fired generator keeps a rising proportion of the total generating capacity in power systems in the last decades, owing to low cost, low carbon emission, and fast response (Qadrdan 2014). As an energy conversion equipment transforming natural gas to electricity power, the gas-fired unit serves as a linkage between the electricity network and gas network. More importantly, with the development of distributed renewable resources and district energy demands (Liu et al. 2016; Wu et al. 2013), the interconnections between

Q.-H. Wu et al., *Large-Scale Integrated Energy Systems*, Energy Systems in Electrical Engineering, https://doi.org/10.1007/978-981-13-6943-8_6

the electricity network and gas network extend from the gas-fired generators to various types of connectivity nodes, such as distributed heating and cooling loads (Gebremedhin 2012; Jing et al. 2014), combined heating and power (CHPs), and combined cooling heating and power (CCHPs) (Wang et al. 2015; Balcombe et al. 2015). Therefore, the interactions between the two networks are playing more and more important roles along with the development of LSIES.

The synergies between the electricity network and gas network have a great influence on the two networks from the viewpoint of economics and security (Gil et al. 2016). From the economics point of view, the gas contracts of price and total generation could affect the unit commitment, economic dispatch, and daily scheduling of the electricity network. From the security point of view, pressure losses, pipeline contingencies, compressor outages, or supply disruptions might lead to forced outages of thermal units or load shedding (Xu et al. 2015). Consequently, the synergies between the electricity network and gas network embedded in the LSIES are worth being investigated for the purpose of economics and security.

Extensive research has conducted on the planning and operation of the combined electricity and gas networks. In Liu et al. (2009), the authors presented the development of a security-based methodology for short term security-constrained unit commitment (SCUC) considering the impact of natural gas transmission constraints. Pricing flexible natural gas supply contracts under uncertainty in hydrothermal markets were discussed in Street et al. (2008). However, these papers considered either the unit commitment (UC) of power system or the pricing of natural gas supply contracts as the master problem rather than treated the electricity network and gas network equally.

The natural gas and electricity optimal power flow was discussed in An et al. (2003). Combined gas and electricity network planning of expansion and energy hubs were presented in Hu et al. (2016) and Salimi et al. (2015), respectively. In Bai et al. (2015), an interval optimization-based operating strategy for gas-electricity-integrated energy systems was proposed. The demand response and wind power uncertainties were taken into consideration. The authors of Zhang et al. (2015) proposed a long-term interdependency of natural gas and electricity infrastructures, incorporating the natural gas transportation planning objective in the co-optimization planning of power generation and transmission systems. In most of the previous research, only the gas-fired generator has been considered as the linkage between the electricity network and gas network (Street et al. 2008; Zhang et al. 2015).

With the increasing of the types of renewable energy resources and energy demands, it is of great importance to extend the synergies of the two networks from the single type to various other types like distributed generators and district heating and cooling loads (Gebremedhin 2012; Jing et al. 2014; Wu et al. 2015). Moreover, as the proportion of gas consumption rise, the gas network plays a significant role as the electricity network from the perspectives of economics and security. Hence, it is wise to treat the gas network and the electricity network coordinately.

This chapter proposes a CSS to optimize conflicting benefits for the daily operation of the electricity and gas networks embedded in the IES. In the combined electricity and gas networks of the IES, we regard both the gas-fired generators and distributed district heating and cooling systems (DHCs) (Zheng et al. 2015) as the interconnections between the two networks. In the case that the two networks are managed by one company operating electricity and gas networks together, the CSS could be achieved without any further effort. In the case that the two networks are managed by two different companies separately, the CSS could be achieved by a joint effort provided by a third party with negotiation between the two companies considering their own business interests and being committed to their social duties. In addition, the linkages of the two networks, gas-fired units, serve as gas consumers of the gas network, but electricity power suppliers of the electricity network, which means that the optimal daily operation of the integrated electricity and gas networks must make a compromise of their own benefits through the synergies.

The CSS consists of a multi-objective optimization procedure and a MADA for the daily operation scheduling of the IES. Based on the multi-optimization model developed in this chapter, the multi-objective optimization procedure is adopted from the multi-objective group search optimizer with adaptive covariance and Lévy flights (MGSO-ACL), which is proved to be efficient for the multi-objective optimization of the IES (Zheng et al. 2015). After the Pareto-optimal set obtained by the multi-objective optimization procedure, the MADA using an IER is utilized to select a final operation solution with adequate evidence fully considering the multiple criteria of the electricity and gas networks and the society interests. In this way, the daily operation of the two networks with conflicting benefits can be tackled with a compromised scheduling solution.

The rest of the chapter is organized as follows: Section 6.2.1 formulates the models of the electricity and gas networks interconnected in the integrated energy system. The coordinated scheduling strategy for the daily operation of the IES is developed in Sect. 6.2.2. Section 6.2.3 carries out the simulation studies to verify the performance of the CSS developed for the daily operation of the IES, and to evaluate the interdependency between the two networks. Finally, the last section draws the conclusion of this chapter.

6.2 Optimal Operation of Integrated Energy Systems with Distributed DHCs Embedded

6.2.1 Integrated Electricity and Gas Networks Modeling

The proposed model of electricity network and gas network integrated into the IES can be illustrated in Fig. 6.1. The system includes electricity transmission system, gas supply system, gas-fired generator, electricity load, gas load, and heating and

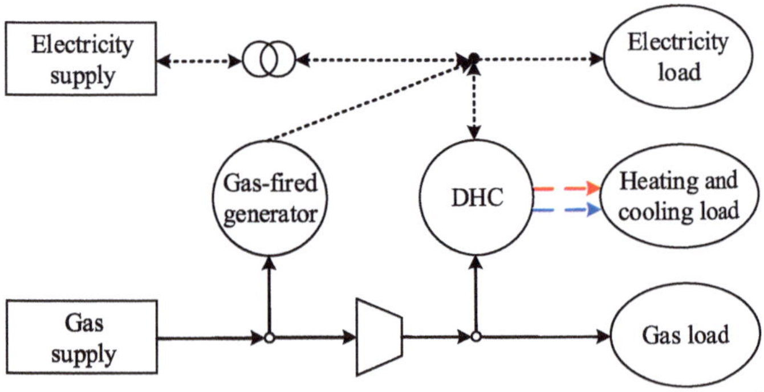

Fig. 6.1 A framework of an IES with electricity and gas networks embedded

cooling load. The electricity network and gas network are closely interconnected by the gas-fired generator and distributed DHC, which can be treated as energy converters between these two energy networks. Noted that the distributed DHCs addressed in this chapter consider the energy input of electricity power and natural gas to serve the heating and cooling load. The models of the electricity network and gas network are described in detail as follows, respectively.

6.2.1.1 Gas Network Model

The gas network is most commonly composed of gas well, gas pipelines, gas compressors, interconnection points, gas storage stations, and gas loads (Qiu et al. 2015; Zhang et al. 2016). In this chapter, gas pipelines, compressors, gas loads, and interconnection points are taken into consideration. The gas well is treated as a constant pressure interconnection point, and it is assumed that the compressors are driven by natural gas. Furthermore, the gas loads include the gas-fired generators and distributed DHCs, apart from the regular gas loads.

(1) Gas pipeline

The gas flow through a pipeline is driven by the pressure difference between the two ends of a pipeline. Several equations have been proposed to compute the gas flow through the pipeline. In this chapter, the most commonly used equation is used to compute the gas flow G in a pipeline between gas network nodes m and n (Mokhatab and Poe 2012; Schroeder et al. 2010; Martínez-Mares and Fuerte-Esquivel 2012):

$$G_{mn} = \text{sgn}(\pi_m, \pi_n)(\frac{77.54T_0}{\pi_0})D_{mn}^{2.5}\sqrt{\frac{|\pi_m{}^2 - \pi_n{}^2|}{L_{mn}\gamma_G T_{mn}^{a} Z^{a} f_{mn}}} \tag{6.1}$$

where π_m and π_n are the pressures at nodes m and n, respectively; $\text{sgn}(\pi_m, \pi_n) = 1$ if $(\pi_m{}^2 - \pi_n{}^2) > 0$ and $\text{sgn}(\pi_m, \pi_n) = -1$ if $(\pi_m{}^2 - \pi_n{}^2) < 0$; T_0 is the base temperature, 520°R; π_0 is the base pressure, 14.65 psia; D_{mn} is the inner diameter of the pipeline between nodes m and n, inch; L_{mn} is the length of pipeline between nodes m and n, miles; γ_G is the gas specific gravity, 0.6; T_{mn}^{a} is the average gas temperature; According to the An et al. (2003), Bai et al. (2015), Zhang et al. (2015), Mokhatab and Poe (2012), Schroeder et al. (2010), the average gas temperature T_{mn}^{a} is practical enough for the model of gas pipelines. Z^{a} is the average gas compressibility factor; and f_{mn} is the friction factor of the pipeline, which strictly depends on the inner diameter of the pipeline and can be given as a function of D_{mn} (Schroeder et al. 2010):

$$f_{mn} = \frac{0.032}{(D_{mn})^{1/3}} \tag{6.2}$$

(2) Gas compressor

During the long transmission distance of gas in pipelines, the gas compressors are installed to compensate for the loss of pressure due to the friction of pipelines. The horsepower of compressor k connected between nodes m and n is mathematically expressed as (An et al. 2003)

$$H_k = \theta_k G_k \left[\left(\frac{\pi_m}{\pi_n} \right)^{Z_k \frac{\delta_k - 1}{\delta_k}} - 1 \right] \tag{6.3}$$

where θ_k is the compressor constant depending on temperature, compressor efficiency and heat ratio; G_k the gas flow at compressor k; Z_k the gas compressibility factor at compressor k; and δ_k the specific heat ratio at compressor k.

The gas compressor must consume horsepower H_k to produce pressure. If the compressor node is coupled with an electricity node, the power will be supplied by the electricity network. In this case, H_k is regarded as an electricity load and will be addressed in the power flow. Otherwise, the compressor will consume natural gas directly from gas flow (Bai et al. 2015). The H_k is denoted as

$$Q(H_k) = a_k H_k{}^2 + b_k H_k + c_k \tag{6.4}$$

where a_k, b_k, and c_k are the coefficients of the gas consumption of the compressor k. In addition, the horsepower H_k of compressor j has to satisfy the physical bound of the compressor.

$$H_k^{\min} \leq H_k \leq H_k^{\max} \tag{6.5}$$

where $H_k^{\min}$ and $H_k^{\max}$ are the minimum and maximum allowed pressure of compressor k, respectively.

The compression ratio between the outlet node m and inlet node n is subject to the following constraint:

$$R_k^{\min} \leq \frac{\max(\pi_m, \pi_n)}{\min(\pi_m, \pi_n)} \leq R_k^{\max} \tag{6.6}$$

where $R_k^{\min}$ and $R_k^{\max}$ are the minimum and maximum allowed compressor ratio, respectively.

(3) Gas loads

From the perspective of the gas network model, the gas-fired generators and distributed DHCs are treated as the gas loads. The gas loads can be regarded as negative gas injections at the gas load nodes. The gas consumption of the gas-fired units is determined by its hourly power generation dispatch of the electricity network (Zhang et al. 2016), which is denoted by

$$G_{\mathrm{GU}i} = \alpha_i + \beta_i P_{\mathrm{GU}i}^t + \gamma_i (P_{\mathrm{GU}i}^t)^2, \quad i \in \mathrm{GU} \tag{6.7}$$

where α_i, β_i, and γ_i are the gas function coefficients of gas-fired unit i, GU is the set of gas-fired units. It is indicated that G_i is considered as a part of the gas loads located in the gas network.

The system structure of the distributed DHCs applied in this chapter is presented in Fig. 6.2. As shown in the figure, the DHC obtains the power input P_{DHC} and gas input G_{DHC} from the electricity network and gas network, respectively. Meanwhile, it serves the heating and cooling loads utilizing the boilers and chillers to transform the power and gas input into heating and cooling demands (Wu et al. 2015). As a consequence, we can get the transformation equations of the DHC, which are shown as follows:

$$\begin{cases} P_{\mathrm{DHC}_1} + P_{\mathrm{DHC}_2} = P_{\mathrm{DHC}} \\ \mathrm{COP}_1 P_{\mathrm{DHC}_1} + \mathrm{COP}_2(\eta_1 P_{\mathrm{DHC}_2} + \eta_2 q_g G_{\mathrm{DHC}} - P_{\mathrm{hd}}) = P_{\mathrm{cd}} \end{cases} \tag{6.8}$$

where COP_1 and COP_2 are the coefficient of performance of the compression chiller and absorption chiller, η_1 and η_2 are the operation efficiency of the electric water boiler and gas-fired water boiler, respectively, and q_g is the calorific value of natural gas.

(4) Gas interconnected point

The gas interconnected point refers to the connected node of the gas network. At a node j of the gas network, the nodal pressure constraint must be satisfied:

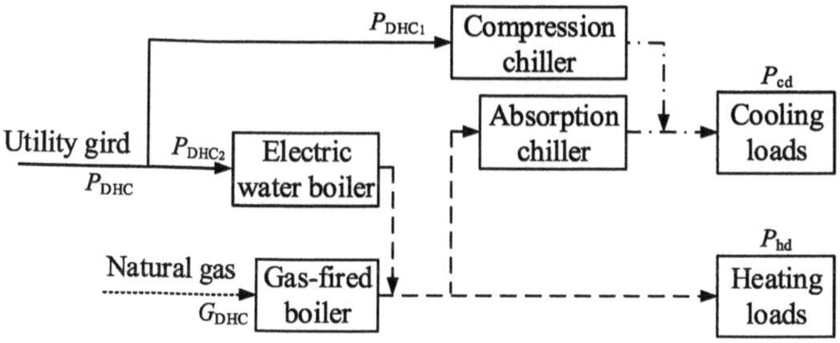

Fig. 6.2 The system structure of distributed DHCs

$$\pi_j^{\min} \leq \pi_j \leq \pi_j^{\max} \tag{6.9}$$

where $\pi_j^{\min}$ and $\pi_j^{\max}$ are the minimum and maximum allowed pressure, respectively.

Specifically, at the gas supply node, the available gas production is limited by the physical characteristics of the gas well, which is given as

$$Q_j^{\min} \leq Q_j \leq Q_j^{\max} \tag{6.10}$$

where $Q_j^{\min}$ and $Q_j^{\max}$ are the minimum and maximum allowed gas supply of gas well j, respectively.

Furthermore, at each node of the gas network, the nodal gas flow balance must be satisfied to assure that the sum of the gas injection to a node is equal to zero.

$$\sum_{m\in n} G_{mn} + \sum_{m\in n} Q_n + \sum_{m\in n} Q_n(H_n) = 0 \tag{6.11}$$

where $\sum_{m\in n}$ is the set of nodes adjacent to node n. Noted that at least one nodal pressure must be considered as a reference to compute all other unknown pressures, and the gas injection computed at this reference point provides the gas flow balance in the gas network.

6.2.1.2 Electricity Network Model

In the model of AC power flow, the active and reactive powers are taken into consideration, the voltage magnitudes and angles of each bus are not constants. Hence, the nonlinear power flow equations have to be tackled during the optimization procedure (An et al. 2003; Martínez-Mares and Fuerte-Esquivel 2012). In this chapter, the electricity network is represented by a direct current (DC) power flow model (Bai et al. 2015). Compared with alternating current (AC) power flow model, the DC formulation is based on the same parameters, but with the following three additional simplifying assumptions: (1) branches can be considered lossless; (2) all bus voltage magnitudes are close to 1 p.u.; (3) voltage angle differences across branches are small enough (Overbye et al. 2004). In this way, the nonlinear power flow equations can be transformed into a linear active power balance equation, without calculating the reactive power. In spite of this, the DC power flow is still applicable for the electricity network modeling proposed in this chapter.

Applying the DC power flow model, the nodal power balance for the electricity network can be expressed in matrix form as

$$I_{\mathrm{G}} P_{\mathrm{G}} + I_{\mathrm{GU}} P_{\mathrm{GU}} - I_{\mathrm{d}} P_{\mathrm{d}} - I_{\mathrm{DHC}} P_{\mathrm{DHC}} = 0 \tag{6.12}$$

where P_{G} and P_{GU} are the power output matrix of thermal units and gas-fired units, respectively; P_{d} and P_{DHC} are the load demand matrix of electricity and distributed DHCs; I_{G}, I_{GU}, I_{d} and I_{DHC} are the incidence matrices of the thermal units, gas-fired units, load demands of electricity, and distributed DHCs, respectively.

Apart from the power flow balance, the power output of a thermal unit has to satisfy its physical limits, which are denoted by

$$P_{\mathrm{G}i}^{\min} \leq P_{\mathrm{G}i} \leq P_{\mathrm{G}i}^{\max} \tag{6.13}$$

$$P_{\mathrm{G}_i}^{t} - P_{\mathrm{G}_i}^{t-1} \leq \mathrm{RU}_{\mathrm{G}i} \tag{6.14}$$

$$P_{\mathrm{G}_i}^{t-1} - P_{\mathrm{G}_i}^{t} \leq \mathrm{RD}_{\mathrm{G}i} \tag{6.15}$$

where $P_{\mathrm{G}i}^{\min}$ and $P_{\mathrm{G}i}^{\max}$ are the minimum and maximum allowed power output of thermal unit i, respectively; $\mathrm{RU}_{\mathrm{G}i}$ and $\mathrm{RD}_{\mathrm{G}i}$ are the maximum ramping up and ramping down rate of the thermal unit Gi.

Spinning reserve is generally supplied by the free capacity of generating units which are able to be activated by the demand of system operators (Partovi et al. 2011). The satisfaction of spinning reserve requirements formulated in (6.16) is significant for both real-time dispatching and daily scheduling in the operation of electricity network.

$$\sum_{i=1}^{N_{\mathrm{G}}} r_{\mathrm{G}i}^{t} \geq r_{\mathrm{D}}^{t} \tag{6.16}$$

where the hourly spinning reserve supply $r_{\mathrm{G}i}^{t}$ is given by (Niknam et al. 2013):

$$\text{Upward: } r_{\mathrm{G}i}^{t} = \min\{(P_{Gi}^{\max} - P_{\mathrm{G}_i}^{t}), \mathrm{RU}_{Gi}\} \tag{6.17}$$

$$\text{Downward: } r_{\mathrm{G}i}^{t} = \min\{(P_{\mathrm{G}_i}^{t} - P_{Gi}^{\min}), \mathrm{RD}_{Gi}\} \tag{6.18}$$

6.2.2 Coordinated Scheduling Strategy

The daily operation optimization for the IES presented in this chapter is related to two different parties: the electricity network and gas network. Even though these two networks simultaneously serve gas load, electricity load, and heating and cooling load, they stand for conflict interests. Therefore, CSS is necessary to maximize the profits of both the electricity network and gas network. The CSS proposed in this chapter consists of a multi-objective optimization procedure and a multiple attribute decision analysis framework, which are presented in detail as follows.

6.2.2.1 Multi-objective Optimization Procedure

Considering the competing benefits of the two networks, a multi-objective optimization model is developed to optimize the operation cost of the electricity network and gas network coordinately. As for the electricity network, the total power generation cost (PGC) includes the fuel cost of thermal units $F_G(P_G)$, start-up SU_G and shut-down cost SD_G (Niknam et al. 2009). The gas supply cost (GSC) of the gas network is the sum of gas production cost Q_j^t minus the benefit of gas consumption Q_k^t. Hence, the objective function of the multi-objective optimization model formulated for the IES considering the competing benefits of the electricity network and gas network is given by

$$\min_X \Big[\sum_t \sum_{i \notin GU} [F_{Gi}(P_{Gi}^t) + SU_{Gi}^t + SD_{Gi}^t], \quad \lambda_g \sum_t \sum_j Q_j^t(\pi_j^t) - \lambda_l \sum_t \sum_k Q_k^t \Big] \tag{6.19}$$

where X is the control variable set: $X = [P, \pi]$, λ_g and λ_l are the corresponding gas prices of gas production and gas consumption, respectively. It should be noted that λ_g and λ_l are normalized during the multi-objective optimization procedure to keep the two objectives in the same number scale.

The corresponding constraints are related to the steady state of both the electricity network and gas network, which are summarized as follows:

Gas network constraints: (1)–(11)
Electricity network constraints: (12)–(18)

In Zheng et al. (2015), the efficiency of multi-objective group search optimizer with adaptive covariance and Lévy flights (MGSO-ACL) is verified for the multi-objective optimization of the IES. Therefore, the MGSO-ACL is addressed to solve the multi-objective optimization model proposed in this chapter. In the adopted MGSO-ACL, two different producers are assigned for the electricity network and gas network, respectively, to search for the optimal direction of the two different objective functions. Meanwhile, the producers are sharing the same scroungers and

Table 6.1 The pseudocode of the MGSO-ACL

```
Set g := 0;
Randomly assign the initial position x_p^(g) and head angles φ^(g) of all members;
Calculate the fitness values of initial members for each objective: F_p(x_p^(g));
WHILE (the termination conditions are not met)
    FOR (each member i in the group)
        Choose producer : Find the member which conferred the best fitness value of each
        objective as producer x_p;
        Perform producing : The producers use a scanning mechanism to randomly sample three
        different directions using (6.20)-(6.22), and get the best direction;
        IF (randn(1)>P(Rangers))
            Perform scrounging : Scroungers Perform scrounging with adaptive covariance
            matrix using (6.25);
        ELSE
            Perform Ranging : Rangers will be dispersed from their current position
            to perform ranging:
            1) Choose a random step size value based on Lévy flights using (6.26);
            2) Move to a new point using (6.27);
        END IF
        Calculate fitness : Evaluate fitness value F_p(x_p^(g)) and run the Pareto-dominance principle;
    END FOR
END WHILE
OUTPUT the best fitness value for each objective F_p(x_best) and the corresponding member x_best.
```

rangers to improve the efficiency of the searching process. The pseudocode of the MGSO-ACL is shown in Table 6.1, and its procedure is described as follows.

(1) Producer searching strategy

In adopted MGSO-ACL, the number of producers is equal to the number of objectives (N_{ob}), which means each producer is assigned to find the best fitness value $F_p(x_p^{(g)}), (p = 1, \ldots, N_{\mathrm{ob}})$ of its corresponding objective. The producers use a scanning mechanism to randomly sample three different directions: straight, left- and right-hand side hypercube, respectively, which are formulated as follows:

$$x_s = x_p^{(g)} + r_1 l_{\max} D_p^{(g)}(\varphi^{(g)}) \tag{6.20}$$

$$x_l = x_p^{(g)} + r_1 l_{\max} D_p^{(g)}(\varphi^{(g)} - r_2 \theta_{\max}/2) \tag{6.21}$$

$$x_r = x_p^{(g)} + r_1 l_{\max} D_p^{(g)}(\varphi^{(g)} + r_2 \theta_{\max}/2) \tag{6.22}$$

where $r_1 \in \mathbb{R}^1$ is a normally distributed random number with mean 0 and standard deviation 1, $r_2 \in \mathbb{R}^{n-1}$ is a uniformly distributed random sequence in the range (0,1), $\varphi_i^{(g)} \in \mathbb{R}^{n-1}$ is the head angle and the unit vector $D(\varphi) \in \mathbb{R}^n$ can be calculated from φ via a polar to Cartesian coordinate transformation (He et al. 2009).

If the best point has a better resource than its current position, then the producer will fly to this point. Otherwise, the producer will stay in its current position and turn its head to a new randomly generated angle:

$$\varphi^{(g+1)} = \varphi^{(g)} + r_2\alpha_{\max} \tag{6.23}$$

where $\alpha_{\max} \in \mathbb{R}^1$ is the maximum turning angle.

On the other hand, if the producer cannot find a better area after a generations, it will turn its head back to zero degree:

$$\varphi^{(g+a)} = \varphi^{(g)} \tag{6.24}$$

where $a \in \mathbb{R}^1$ is a constant.

(2) Scroungers' behaviors with adaptive covariance

The adaptive covariance matrix, obtained by cumulatively learning for the information organized from the group members of each generation, is employed to get a reliable estimator for determining the evolution path and step size for scroungers' behaviors.

The offspring of kth scrounger, $x_k^{(g+1)}$, can be modeled as follows (Auger and Hansen 2012):

$$x_k^{(g+1)} = m^{(g)} + \sigma^{(g)}\mathcal{N}(0, C^{(g)}) \quad k = 1, \ldots, \lambda \tag{6.25}$$

where $\mathcal{N}(0, \mathrm{I})$ means a multivariate normal distribution with zero mean and unity covariance matrix, $\sigma > 0$ is the step size, λ is the number of the scroungers, superscript g denotes the generation number, ($g = 0, 1, 2, \ldots$), and n is the dimension of the function.

Mean vector $m^{(g)}$ of the searching distribution is a weighted average of μ successful individuals selected from the sample $x_1^{(g)}, \ldots, x_\lambda^{(g)}$. Covariance matrix C is updated based on mean vector, and the evolution path and step size are accordingly determined by the covariance matrix (Auger and Hansen 2012).

(3) Rangers' walks

In this chapter, Lévy flights (Yang 2010) are introduced as rangers' searching technique rather than the random walks. The step size value of the ith ranger is chosen randomly as follows:

$$s_i = 0.01 \left(\frac{u_i}{v_i}\right)^{1/\beta} (x_i^{(g)} - x_p^{(g)}) \tag{6.26}$$

where $u = \phi\,\mathrm{randn}(n)$, $v = \mathrm{randn}(n)$, $\beta = 1.50$, and n is the number of variables. The $\mathrm{randn}(n)$ function generates a uniform integer between $[1, n]$, and the ϕ is computed by

$$\phi = \left(\frac{\Gamma(1+\beta)\sin(\pi\beta/2)}{\Gamma((1+\beta)/2) \cdot \beta \cdot 2^{(\beta-1)/2}} \right)^{1/\beta}$$

where Γ denotes the *gamma function*.

Consequently, rangers will move to the new point following the direction as

$$x_i^{(g+1)} = x_i^{(g)} + \text{randn}(n)s_i \tag{6.27}$$

In this way, the individuals, x_i, of the MGSO-ACL are updated according to the fitness value of the multiple objectives. Further details about the MGSO-ACL can be referred to Zheng et al. (2015). Applying the MGSO-ACL, a Pareto-optimal set for the daily operation of the IES can be obtained.

6.2.3 Simulation Studies

6.2.3.1 System Description

In this section, simulation studies conducted on a test-integrated energy system with electricity network and gas network embedded are carried out to figure out coordinated scheduling strategy for the daily operation. The test IES is adopted by a modified IEEE 30-bus system and a 15-node gas network, which is shown in Fig. 6.3. The detailed parameter data can be found in An et al. (2003), Martínez-Mares and Fuerte-Esquivel (2012).

As presented in the figure, two gas-fired units are connected to the electricity network at buses 7 and 12, respectively. Moreover, three distributed DHCs are located at buses 8, 11, 21, and 30 of the electricity network. The gas network consists of 15 nodes, which contains 2 gas source nodes (1,2), 2 gas load nodes (3,13), 2 nodes connected to gas-fired units (8,15) and 3 nodes connected to distributed DHCs (4,9,14). We can see from the figure that the electricity network is interconnected with the gas network through the two gas-fired generators and three distributed DHCs. The gas-fired units serve as gas loads in the gas network, and they are also power sources for the electricity network. The distributed DHCs also serve as gas loads in the gas network, and they can also exchange power with the electricity network.

The daily curves of electricity load, gas load, and heating and cooling load are summarized in Fig. 6.4. Noted that the gas load shown in the figure only includes the gas demand in gas load nodes (3,13) of the gas network, without the gas demands of gas-fired units and distributed DHCs. Moreover, the heating and cooling load demands are satisfied by the power from distributed DHCs.

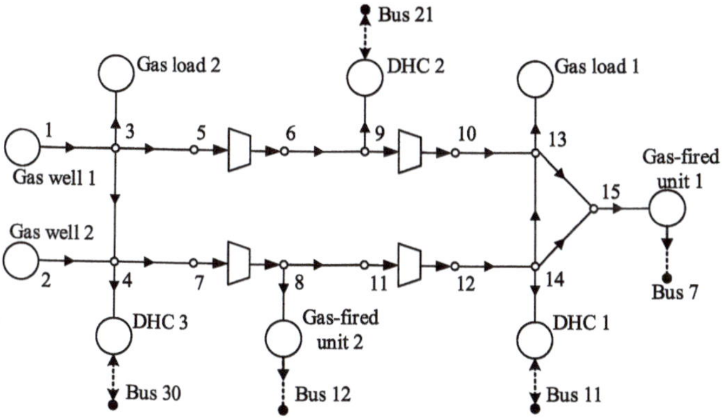

Fig. 6.3 The 15-node gas network coupled with the IEEE 30-bus network of the test IES

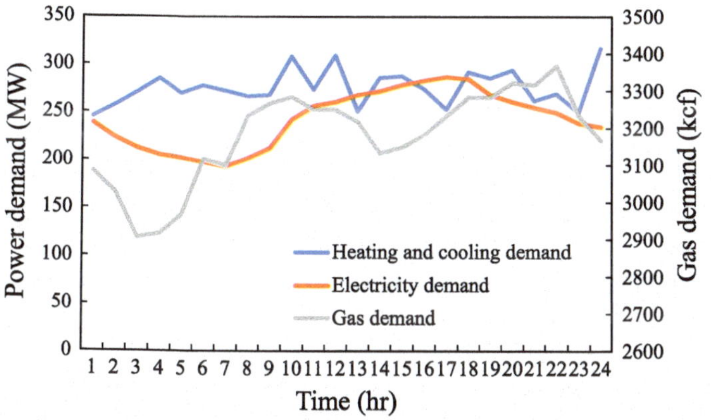

Fig. 6.4 The daily curves of electricity load, gas load, and heating and cooling load

6.2.3.2 Results Comparisons

The CSS proposed in this chapter is utilized to obtain the optimal daily operation solution for the test IES. The multi-objective optimization algorithm, MGSO-ACL, is evaluated in 100 independent runs, with comparisons with the non-dominated sorting genetic algorithm II (NSGA-II) (Deb et al. 2002). The simulation results of the PGC of the electricity network and the GSC of the gas network at $t = 1$ is shown in Fig. 6.5. As presented in the figure, the operation costs of the electricity network and gas network are in conflict with each other, and they compromise with each other in order to reach both of their optimal solutions. The results imply that it is necessary to optimize the two dependent interests coordinately, rather than simply optimizing the sum of the two conflicting interests using single-objective optimization method.

For the metrics comparisons between MGSO-ACL and NSGA-II, the index of hypervolume (HV), the mean Euclidian distance (MED), the spacing index (SI), the

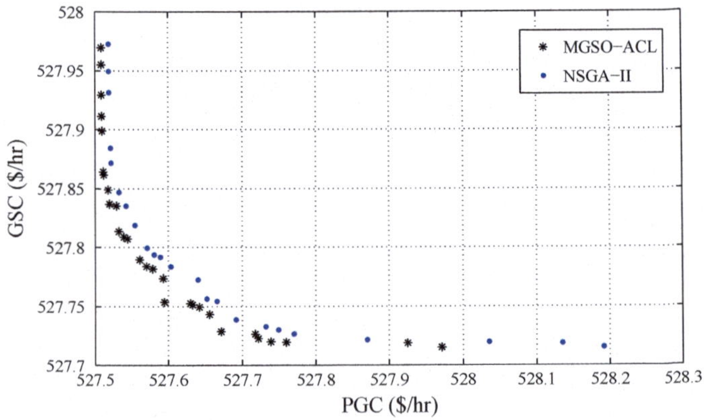

Fig. 6.5 Pareto-optimal fronts of MGSO-ACL and NSGA-II for the test IES at $t = 1$

Table 6.2 Metrics comparisons of Pareto-optimal fronts obtained by MGSO-ACL and NSGA-II

Algorithm	NPS	HV	MED	SI	CT (s)
MGSO-ACL	29	0.0595	0.1874	0.0231	263.70
NSGA-II	23	0.0519	0.2607	0.0377	306.51

number of Pareto-optimal solutions (NPS), and computation time (CT) is addressed to evaluate the quality (convergence and diversity) of the Pareto-optimal fronts. The reference points for HV and MED are set as $(527.9, 527.9)$ and $(527.5, 527.7)$, respectively.

According to the metrics comparisons shown in Table 6.2, MGSO-ACL can find more Pareto-optimal solutions than NSGA-II under the same number of independent runs. Meanwhile, the values of HV and MED show that the Pareto-optimal fronts obtained by MGSO-ACL have better quality of convergence and diversity than those obtained by NSGA-II. In addition, SI regarding to MGSO-ACL is smaller, which means that the Pareto-optimal solutions obtained by MGSO-ACL are distributed more evenly. In addition, it can be seen from the table that the computation time of MGSO-ACL is less than that of NSGA-II, which means MGSO-ACL is more efficient. Therefore, it is demonstrated that MGSO-ACL finds a better set of Pareto-optimal solutions, compared with NSGA-II.

In order to select a final operation solution for the test IES, the proposed MADA method, IER, is applied to generate the overall assessments of all the alternative solutions. As a consequence, we can get the ranking of all the solutions in the Pareto-optimal set, which is shown in Table 6.3.

As shown in the table, the optimal solutions of the electricity network and gas network are making compromise with each other. The objective functions f_{ele} and f_{gas} cannot reach the minimum solution simultaneously. Consequently, the MADA-based IER should be applied to select a final operation solution with adequate evidence fully

Table 6.3 The ranking of the solutions in the Pareto-optimal fronts

No.	1	2	3	4	5	6	7	8	9	10
PGC ($/hr)	527.5086	527.5088	527.5089	527.5092	527.5098	527.5113	527.5121	527.5183	527.5200	527.5296
GSC ($/hr)	527.9695	527.9551	527.9292	527.9112	527.8988	527.8642	527.8616	527.8489	527.8368	527.8353
u_{min}	0.605	0.612	0.655	0.661	0.675	0.690	0.712	0.788	0.797	0.815
u_{max}	0.481	0.480	0.498	0.554	0.572	0.627	0.662	0.733	0.782	0.772
u_{avg}	0.543	0.546	0.577	0.607	0.624	0.659	0.687	0.761	0.790	0.793
Ranking	27	26	25	22	21	18	17	15	14	13
$P(m > n)$ (%)	80.46	51.17	60.55	54.43	57.62	57.58	75.22	100	91.43	56.90
No.	11	12	13	14	15	16	17	18	19	20
PGC ($/hr)	527.5328	527.5397	527.5448	527.5613	527.5707	527.5791	527.5931	527.5948	527.6294	527.6324
GSC ($/hr)	527.8133	527.8086	527.8071	527.7894	527.7837	527.7817	527.7736	527.7535	527.7525	527.7514
u_{min}	0.814	0.850	0.841	0.887	0.902	0.889	0.921	0.938	0.922	0.901
u_{max}	0.801	0.774	0.792	0.812	0.823	0.862	0.894	0.906	0.883	0.882
u_{avg}	0.808	0.812	0.817	0.845	0.863	0.876	0.908	0,922	0.903	0.892
Ranking	12	11	10	8	7	6	2	1	3	4
$P(m > n)$ (%)	75.00	55.06	53.60	56.00	58.44	62.26	57.58	74.58	68.97	57.69
No.	21	22	23	24	25	26	27	28	29	–
PGC ($/hr)	527.6415	527.6554	527.6713	527.7177	527.7218	527.7391	527.7579	527.9251	527.9714	–
GSC ($/hr)	527.7494	527.7433	527.7288	527.7266	527.7229	527.7202	527.7196	527.7187	527.7152	–
u_{min}	0.904	0.878	0.711	0.688	0.675	0.662	0.651	0.523	0.520	–
u_{max}	0.871	0.803	0.685	0.603	0.599	0.532	0.512	0.432	0.411	–
u_{avg}	0.888	0.841	0.698	0.646	0.637	0.597	0.582	0.473	0.466	–
Ranking	5	9	16	19	20	23	24	28	29	–
$P(m > n)$ (%)	70.00	69.35	64.47	55.27	57.54	55.76	51.69	56.00	–	–

considering both the multiple objectives and the multiple criteria of the electricity network and gas network.

The set of basic attributes is $E = \{$pressure deviation (PD), polution emission $(PM)\}$, which are reliability index of the gas network, and the environmentally friendly index of the thermal units in the electricity network:

$$PD = \sum |\pi_i - \pi_{\text{ref}}|, \ PM = \sum (\tau_{0j} + \tau_{1j} P_{\text{G}j} + \tau_{1j} P_{\text{G}j}^2)$$

where i and j are the indexes of nodes in gas network and thermal units; π_{ref} is the reference pressure of each node in the gas network; τ_{0j}, τ_{1j} and τ_{2j} are the pollution emission coefficients of thermal unit j. The values of PD and PM will be figured out after the optimization procedure is finished. The set of corresponding relative weights for the attributes is set as the same in this chapter, $\omega = [0.5, 0.5]$. In addition, the set of evaluation individual grades is set as $H_{pp} = \{$poor, indifference, average, good, excellent$\}$, and then the interval grades H_{pq} are generated between the individual grades.

Applying the procedure of the IER, the utility evaluation of each solution given by the maximum, minimum and average expected utility evaluation is summarized in Table 6.3. Accordingly, based on the utility evaluation, we can get the ranking of the solutions in the Pareto-optimal fronts, assigning with the superior percentage $P(m > n)$ for the neighboring ranking m and n. For example, the 18th solution ranks first with 74.58% more superior than the solution 17th ranking in the second place. Accordingly, the 18th can be selected as the final operation point at this moment.

Applying the CSS over the 24-hr scheduling period, we can obtain the daily operation solutions for the test IES. The physical meaning of the solution obtained by the CSS method is the optimal daily operation point for the test IES, considering the conflicting benefits of the electricity and gas networks and satisfying the operation constraints. The daily operation point includes the power generation scheduling of the thermal units, gas production scheduling of the gas wells, the power flow of the electricity network and the gas flow of the gas network. The power generation scheduling of the electricity network and the gas production scheduling of the gas network are shown in Fig. 6.6 and Fig. 6.7, respectively.

As shown in Fig. 6.6, the optimal power outputs of the thermal units are related to their generation costs to minimize the PGC of the electricity network. Hence the power output of unit 1 which is with lower generation cost are higher than the other units. Furthermore, it can be seen from the figure that the power outputs of units 2, 3, 4 are steady over the scheduling period, while the output of unit 1 fluctuates along with the trend of the electricity load demand. This is because unit 1 is regarded as the reference bus to balance the power demand. According to Fig. 6.7, the sum of gas production in gas well 1 and gas well 2 keeps in a stable trend, and the gas production of gas well 2 is lower than that of gas well 1 for its lower price to reduce the GSC of the gas network. Compared to the load demand shown in Fig. 6.4, we can see that the sum of gas production is much larger than the amount of gas load demand. The rest of the gas production is utilized to serve the gas-fired units and distributed DHCs.

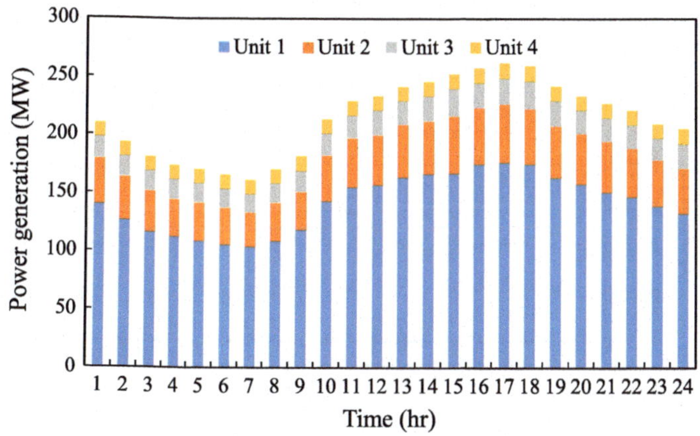

Fig. 6.6 The optimal power generation scheduling of the thermal units

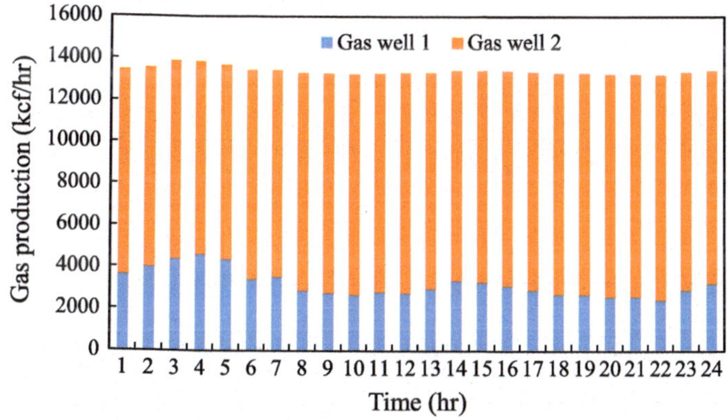

Fig. 6.7 The optimal gas production scheduling of the gas wells

Moreover, the amount of electricity load is larger than the sum of power generation, and the rest of the part is served by the gas-fired units or the distributed DHCs.

The PGC of the electricity network, the GSC of the gas network and the total cost (TC) over the 24-hr are summarized in Table 6.4, and compared with optimizing the sum of the two conflicting interests using single-objective optimization method (SOO). The results show that in the SOO method, not only the PGC or the GSC cannot reach an optimal result, the TC of the two competing benefits is worse than that of the CSS, either. From the results presented in Table 6.5, we can see that compared with those of the SOO, the daily cost savings of the CSS for the electricity network and gas network are \$19.32 and \$27.13, respectively. Therefore, it is necessary to treat the conflicting benefits of the two networks as two objectives using the proposed CSS for the purpose of economics.

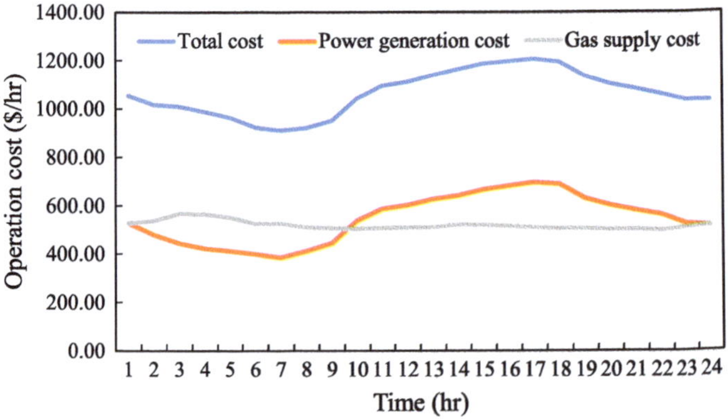

Fig. 6.8 The optimal operation cost for the test IES over the 24-hr

In addition, the optimal operation costs of the electricity network, gas network and the total cost are depicted in Fig. 6.8. Compared to the trend of cost solution with the load demand shown in Fig. 6.4, the PGC of the electricity network is related with the electricity load demand, while the GSC of the gas network do not vary along with the variation of the gas load demand. This is because the gas load occupies only a part of the total gas supply, instead, the gas-fired units and distributed DHCs take the rest of the gas load demand.

6.2.3.3 Interdependency Analysis of the Electricity and Gas Networks

(1) Scenario 1: Variations of the electricity, gas, and heating and cooling loads

Furthermore, in order to evaluate the interdependency between the electricity network and gas network, the variations PGC and GSC via the following variations of the electricity, gas and heating and cooling loads are analyzed: (1) keeping the electricity load and heating and cooling loads unchanged, the gas load varies from −25% to 25% by a step of 5%; (2) keeping the gas load and heating and cooling loads unchanged, the electricity load varies from −25% to 25% by a step of 5%; (3) keeping the gas load and electricity load unchanged, the heating and cooling loads vary from −25% to 25% by a step of 5%. The corresponding variation of the PGC and GSC of the two networks are shown in Fig. 6.9, Fig. 6.10, and Fig. 6.11, respectively.

As shown in Fig. 6.9, the PGC randomly fluctuates less than $1.0 as the gas load varies from −25% to 25%, while the GSC increases along with the gas load. According to Fig. 6.10, we can see that as the electricity load rises, the PGC of the electricity network increases almost linearly with a high slope, while the GSC of the gas network stays steadily within an error of $1.5. By contrast, as the heating and cooling load varies from −25% to 25% shownin Fig. 6.11, both of the PGC and GSC

Table 6.4 The comparisons of daily operation costs obtained by the CSS and SOO

Time		1	2	3	4	5	6	7	8
CSS	PGC	527.59	479.99	444.60	423.68	414.25	400.89	387.74	413.65
	GSC	527.75	538.08	566.80	565.63	551.32	524.68	525.42	509.87
	TC	1055.34	1018.07	1011.40	989.31	965.57	925.57	913.16	923.52
SOO	PGC	528.01	482.14	445.33	424.13	415.14	401.98	387.43	413.25
	GSC	529.41	538.94	569.38	565.80	552.51	523.96	526.37	519.73
	TC	1057.42	1021.08	1014.71	989.93	967.65	925.94	913.80	923.98
Time		9	10	11	12	13	14	15	16
CSS	PGC	447.02	541.45	590.29	605.39	630.32	644.96	669.35	684.60
	GSC	506.41	503.99	507.21	508.85	510.98	521.01	519.22	514.78
	TC	953.43	1045.44	1097.50	1114.24	1141.30	1165.97	1188.57	1199.38
SOO	PGC	447.28	541.71	590.28	606.00	631.62	645.51	670.56	686.15
	GSC	507.07	505.26	508.63	509.10	512.36	521.55	520.45	515.45
	TC	954.35	1046.97	1098.91	1115.10	1143.98	1167.06	1191.01	1201.60
Time		17	18	19	20	21	22	23	24
CSS	PGC	698.78	691.51	631.45	604.04	583.85	565.44	528.03	517.68
	GSC	509.23	505.03	505.15	500.88	501.32	496.58	510.06	518.59
	TC	1208.01	1196.54	1136.60	1104.92	1085.17	1062.02	1038.09	1036.27
SOO	PGC	699.92	692.21	632.68	605.23	585.66	566.46	529.18	517.98
	GSC	509.60	506.13	505.56	501.03	501.83	496.88	509.95	519.00
	TC	1209.52	1198.34	1138.24	1106.26	1087.49	1063.34	1039.13	1036.98

Table 6.5 The comparisons of daily total costs obtained by the CSS and SOO

Method	PGC ($)	GSC ($)	TC ($)
SOO	13145.85	12475.97	25621.82
CSS	13126.53	12448.83	25575.36
Cost savings ($/day)	19.32	27.13	46.45

tend to increase by different ramp rates, which means that the heating and cooling load has an impact on the electricity network and gas network.

These interdependency analyses imply that the gas load only impacts the gas network, and the electricity load only has a great influence on the electricity network. Moreover, as the interconnected point via the distributed DHCs between the electricity and gas networks, the heating and cooling load can affect the two networks simultaneously. Therefore, the interdependency of the electricity network and gas network relies on the variations of the heating and cooling load, rather than the electricity load or the gas load.

(2) Scenario 2: Variations on the numbers of the electric power driven compressors

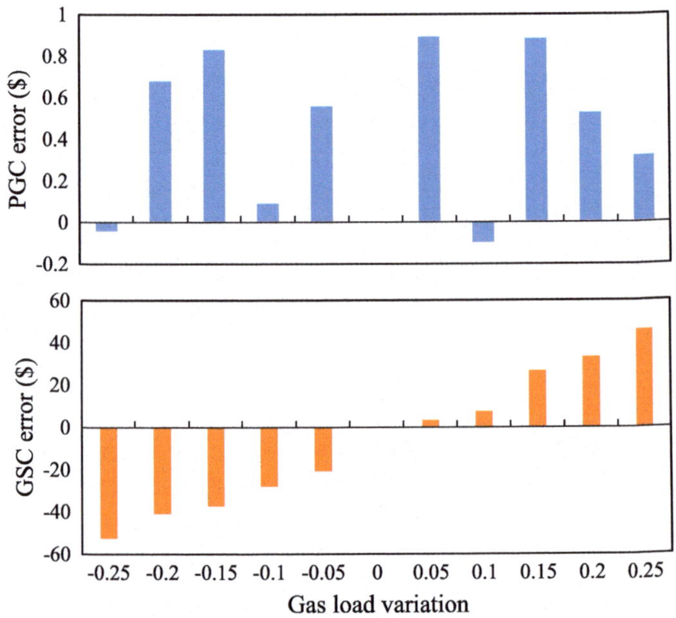

Fig. 6.9 The operation cost error via the variation of gas load

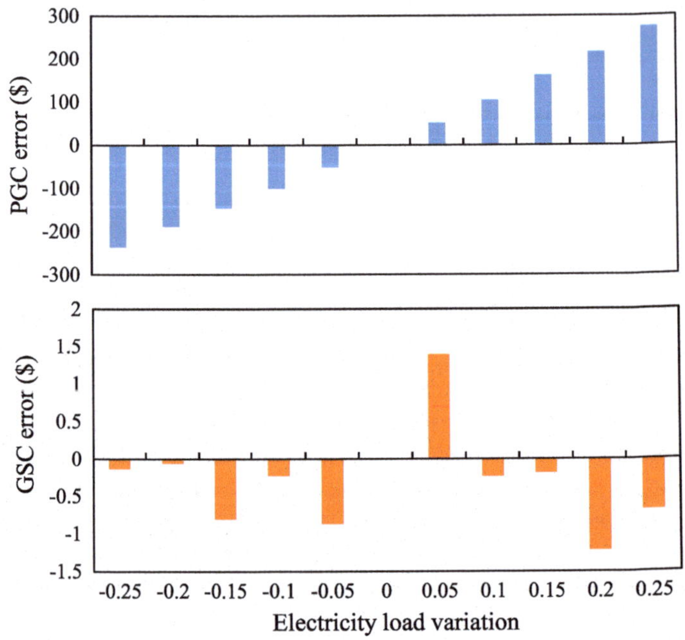

Fig. 6.10 The operation cost error via the variation of electricity load

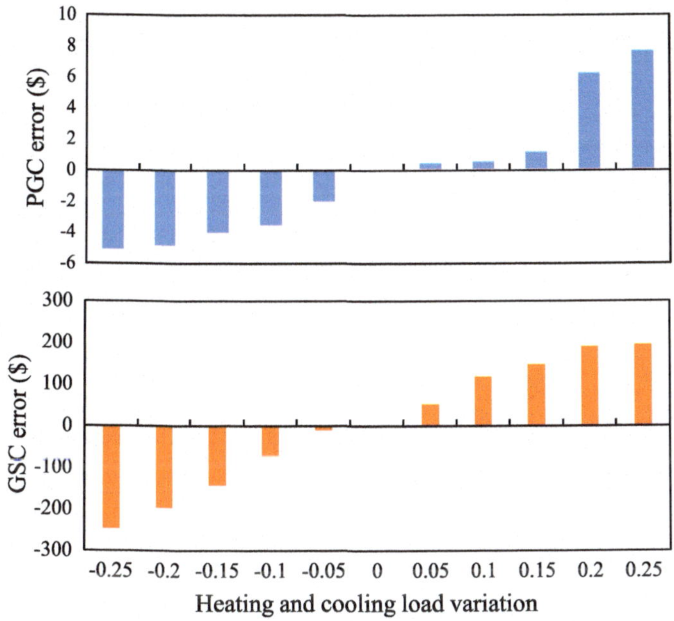

Fig. 6.11 The operation cost error via the variation of heating and cooling load

According to the model of gas compressor in (6.3)–(6.6), the compressor will enhance the interdependency between the electricity and gas networks if it is coupled with an electricity node. In this case, the input horsepower of the compressor is regarded as an electricity load.

In order to further evaluate the interdependency between the electricity and gas networks via the variation of the number of electricity-driven compressor, we carry out the following experiments: (1) keeping the electricity load and heating and cooling loads unchanged, the gas load varies from −25% to 25% by a step of 5% with the number of electricity-driven compressor $N_c = 0$, $N_c = 2$ and $N_c = 4$; (2) keeping the gas load and heating and cooling loads unchanged, the electricity load varies from −25% to 25% by a step of 5% with the number of electricity-driven compressor $N_c = 0$, $N_c = 2$ and $N_c = 4$. The corresponding results are shown in Fig. 6.12 and Fig. 6.13, respectively.

According to the results, the number of the electricity-driven compressor will strengthen the relationship between the PGC and the variation of the gas load. This is because the power consumption of the compressor is related to the gas flow over the compressor branch. The more gas load demand, the more gas flow needed. Hence the power consumption of the compressor increase, resulting in the rise of the PGC. By contrast, the number of the electricity-driven compressor has little impact on the interdependency between the GSC and the variation of the electricity load.

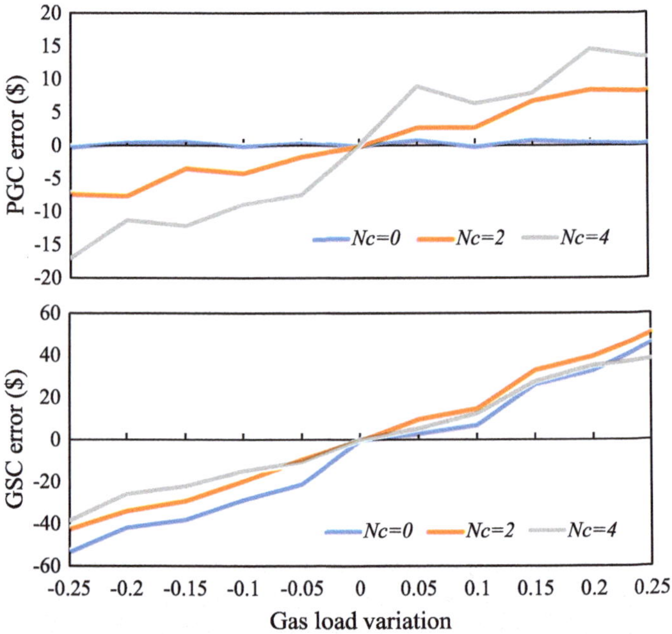

Fig. 6.12 The operation cost error via the variation of gas load under different numbers of electricity-driven compressor

6.3 Coordinated Scheduling Strategy for LSIES Considering Conflict Benefits

6.3.1 Integrated Energy System Modeling

Figure 6.14 shows the proposed IES with multiple DENs embedded. The IES is composed of an electricity utility network, a natural gas utility network, and a number of DENs with multiple DESs and multiple EUs. The scenario of DES set in this section is in summer to supply E&C energy to EUs. In each DEN, a number of DESs perform as local energy suppliers purchasing E&G for generating and selling E&C energy to EUs who determine their energy demands according to energy prices, but natural gas cannot be directly transmitted from the DES to EUs in the IES. Considering the fact that the distance between node to node in E&G networks is actually too long to achieve non-electricity energy transmission, E&C energy demands of EUs are in a perfectly satiated state thanks to the local DESs in each DEN.

In the latter part of this section, the mathematical models of electricity network, gas network, and DENs with multiple DESs and EUs are described.

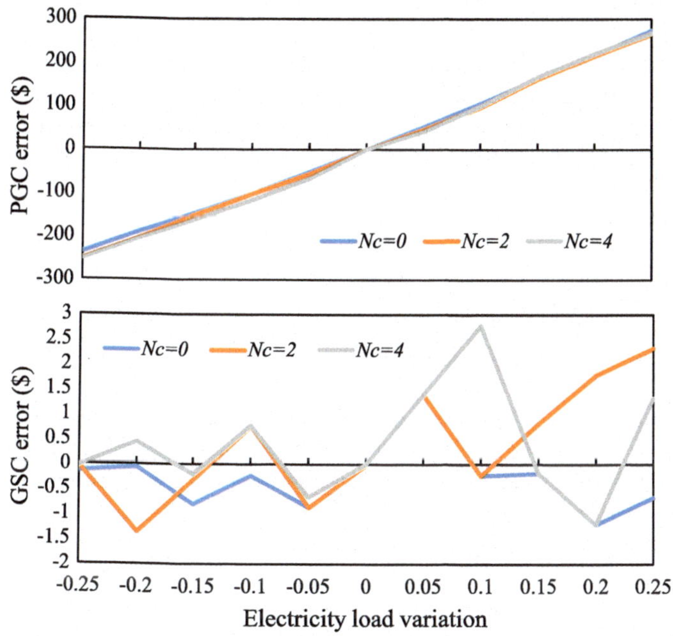

Fig. 6.13 The operation cost error via the variation of electricity load under different numbers of electricity-driven compressor

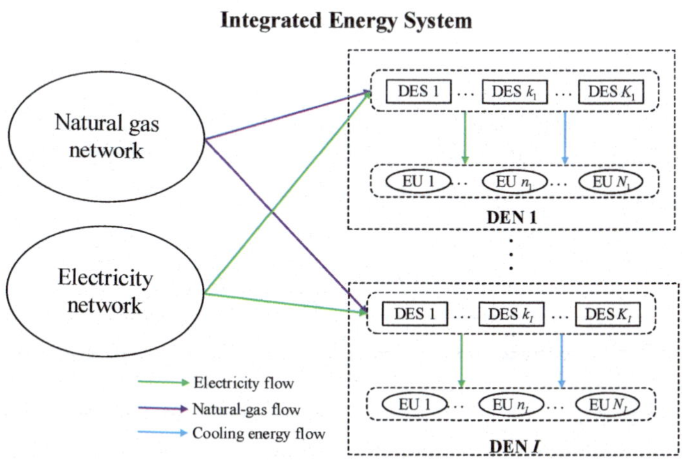

Fig. 6.14 Framework of an IES with multiple DENs embedded

6.3.1.1 Electricity Network Model

Optimizing the generation dispatch of the generators has a great effect on the optimal operation of electricity networks, and this section considers a set of constraints from the viewpoint of stable operation, power balance, and demand response (Li et al. 2016) which can be formulated as follows:

$$\min \ J_e(U_e, X_e) = \sum_{\varepsilon=1}^{N_G} [\alpha_\varepsilon^e (P_{G_\varepsilon})^2 + \beta_\varepsilon^e P_{G_\varepsilon} + \gamma_\varepsilon^e] \tag{6.28}$$

$$\text{s.t.}\ g_e(U_e, X_e) = 0 \tag{6.29}$$

$$h_e(U_e, X_e) \le 0 \tag{6.30}$$

where J_e indicates the objective function which is usually the total operating cost of generators. P_{G_ε} is the real power output of generator ε. α_ε^e, β_ε^e and γ_ε^e are the corresponding coefficients of fuel cost. N_G is the number of generators. Moreover, U_e and X_e are the vectors of control variables and state variables in electricity network, respectively. g_e and h_e are equality and inequality constraints of electricity network, respectively.

The vector of control variables U_e includes active outputs and voltages of generators, transformer tap ratios and reactive power generation of shunt devices, which can be presented as follows:

$$\begin{aligned} U_e{}^T = [&P_{G_2}, \ldots, P_{G_{NG}}, V_{G_1}, \ldots, V_{G_{NG}}, \\ &T_1, \ldots, T_{N_T}, Q_{C_1}, \ldots, Q_{C_{NC}}] \end{aligned} \tag{6.31}$$

where N_T and N_C are the total numbers of transformer tap rations and shunt devices, respectively. In addition, the vector of state variables X_e in electricity network includes active power of slack bus, voltages of load buses, generator reactive powers and apparent power flows, which can be presented as follows:

$$\begin{aligned} X_e{}^T = [&P_{G_1}, V_{L_1}, \ldots, V_{L_{ND}}, \\ &Q_{G_1}, \ldots, Q_{G_{NG}}, S_1, \ldots, S_{N_E}] \end{aligned} \tag{6.32}$$

where N_D and N_E are the total numbers of load buses, and power network branches, respectively.

The equality constraints $g_e(U_e, X_e)$ represent the balance bounds of active and reactive power. Moreover, the inequality constraints $h_e(U_e, X_e)$ involve bounds of active and reactive power outputs of generators, transformer tap ratios, reactive power generation of shunt devices, voltages of buses, and power flows of transmission lines. The formulations of g_e and h_e in detail can be referred to Zheng et al. (2015).

6.3.1.2 Natural Gas Network Model

The natural gas transmission system commonly consists of gas wells, pipelines, compressors and loads. Natural gas is injected from the gas wells and transmitted through gas pipelines to gas loads. Noted that the optimization of the steady-state natural gas problem takes the amount of gas infusion of gas wells (all but that of slack node) and the compression ratios as controllable variables and determines the state variables including gas flow in pipelines, nodal pressures (apart from slack node) and the amount of gas infusion of slack node (Zhang et al. 2016).

(a) Pipeline model

The amount of gas flow between node ω and υ in a pipeline can be calculated by (An et al. 2003)

$$f_{\omega\upsilon} = \rho(\pi_\omega, \pi_\upsilon) C_{\omega\upsilon} \sqrt{\mid \pi_\omega^2 - \pi_\upsilon^2 \mid} \tag{6.33}$$

$$\rho(\pi_\omega, \pi_\upsilon) = \begin{cases} 1 & \pi_\omega \geq \pi_\upsilon \\ -1 & \pi_\omega < \pi_\upsilon \end{cases} \tag{6.34}$$

where $f_{\omega\upsilon}$ represents the gas flow between node ω and υ. π_ω and π_υ denote pressures at node ω and υ, respectively. $C_{\omega\upsilon}$ is the pipeline constant associated with length, diameter, friction, temperature, and compressibility.

(b) Compressor model

Compressors plays a crucial role in the natural gas network, like compensating the full of pressure of pipelines and holding the level of pressure within a certain range. The natural gas flow through compressor τ, which locates between node ω and υ, can be given as follows (Liu et al. 2009):

$$f_{\omega\upsilon} = \rho(\pi_\omega, \pi_\upsilon) \frac{H_\tau}{\mu_1 - \mu_2 (R_\tau)^{\mu_3}} \tag{6.35}$$

where μ_1, μ_2, and μ_3 are parameters corresponding to the compressor design. H_τ and R_τ represent the horsepower consumption and compression ratio of compressor τ, respectively.

In this section, it is supposed that the power of the compressor is derived from gas turbine, by which the amount of gas consumed can be expressed as

$$\xi_\tau = \alpha_\tau^c + \beta_\tau^c H_\tau + \gamma_\tau^c (H_\tau)^2 \tag{6.36}$$

where ξ_τ denotes the amount of gas consumed by compressor τ. α_τ^c, β_τ^c, and γ_τ^c are consumption coefficients of compressor τ, respectively.

(c) Optimal operation of gas network

The optimal operation of gas network relates to the optimization of the gas dispatch of gas wells considering a set of constraints, like stable operation, gas flow balance, and demand response (De Wolf and Smeers 2000), which can be depicted as

$$\min\ J_{\mathrm{g}}(U_{\mathrm{g}}, X_{\mathrm{g}}) = \sum_{\kappa=1}^{N_{\mathrm{GW}}} [c_{\kappa}^{\mathrm{g}}(F_{\kappa})^2 + b_{\kappa}^{\mathrm{g}} F_{\kappa} + a_{\kappa}^{\mathrm{g}}] \tag{6.37}$$

$$\text{s.t. } g_{\mathrm{g}}(U_{\mathrm{g}}, X_{\mathrm{g}}) = 0 \tag{6.38}$$

$$h_{\mathrm{g}}(U_{\mathrm{g}}, X_{\mathrm{g}}) \leq 0 \tag{6.39}$$

where J_{g}, F_{κ}, a_{κ}^{g}, b_{κ}^{g}, c_{κ}^{g}, N_{GW}, U_{g}, X_{g}, g_{g}, and h_{g} indicate the same mathematical implications as Eq. (6.28), the difference is that the physical objects described here is gas wells of natural gas network.

U_{g} includes the amount of gas infusion of gas wells all but that of slack node and compression ratios. In addition, X_{g} consists of gas infusion of slack node, nodal pressures all but slack node and pipeline gas flows.

The equality constraints of gas network $g_{\mathrm{g}}(U_{\mathrm{g}}, X_{\mathrm{g}})$ are the balance bounds of gas flow, which must be satisfied to guarantee that the sum of the gas input and the sum of the gas output are consistent (Qiao et al. 2016).

$$A_{\mathrm{GW}} F_{\kappa} - A_{\mathrm{P}} f_{\omega \upsilon} - A_{\mathrm{C}} \xi_{\tau} - A_{\mathrm{L}} L_l^{\mathrm{g}} = 0 \tag{6.40}$$

where A_{GW}, A_{P}, A_{C}, and A_{L} describe the incidence matrices of gas wells, pipelines, compressors, and gas loads, respectively. L_l^{g} denotes the amount of gas load of load l.

There is also a set of inequality constraints $h_{\mathrm{g}}(U_{\mathrm{g}}, X_{\mathrm{g}})$ demonstrating the physical bounds of devices in gas network. The corresponding bounds of the amount of gas infusion of gas wells, compression ratios, and nodal pressures can be described as follows:

$$F_{\kappa,\min} \leq F_{\kappa} \leq F_{\kappa,\max} \tag{6.41}$$

$$R_{\tau,\min} \leq \frac{\max(\pi_{\omega}, \pi_{\upsilon})}{\min(\pi_{\omega}, \pi_{\upsilon})} \leq R_{\tau,\max} \tag{6.42}$$

$$\pi_{\omega,\min} \leq \pi_{\omega} \leq \pi_{\omega,\max} \tag{6.43}$$

6.3.1.3 District Energy Network

In the proposed IES, a number of I DENs, $\tilde{\mathrm{I}} = \{1, 2, \ldots, I\}$, are considered, each of which serves energy transmission infrastructure to K DESs, $\Bbbk_i = \{1, 2, \ldots, K_i\}$, and N EUs, $\aleph_i = \{1, 2, \ldots, N_i\}$ locating in the region it covered. Figure 6.15 describes the illustrative diagram of the DEN. In each DEN, DESs will behave competitively as energy suppliers, and therefore a localized competitive market forms. In another word, each DES will compete with his rivals and trades with EUs because of its

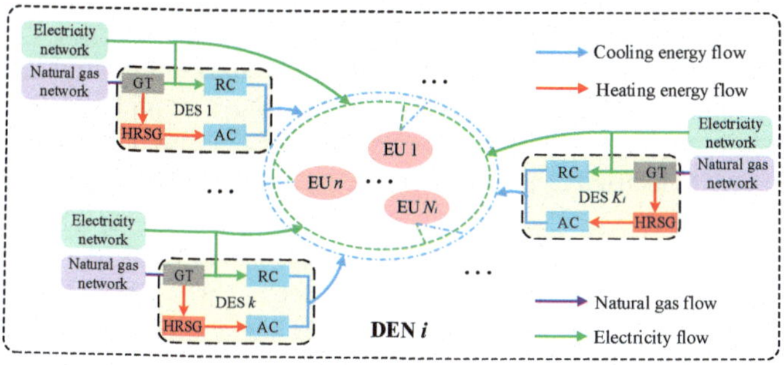

Fig. 6.15 Illustrative diagram of district energy network

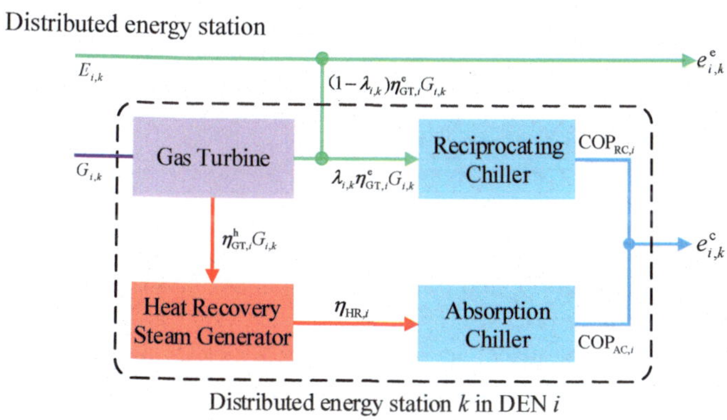

Fig. 6.16 Illustrative diagram of the distributed energy station

own benefit-seeking psychology. In this section, the E&C energy markets in each DEN are assumed to be perfectly competitive so that all DESs would charge EUs the uniform unit energy prices. In this section, the mathematical models of DES and EU will be discussed. For the purpose of illustration, we will focus on describing the DESs and EUs in one DEN, such as DEN i, $\forall i \in \tilde{\mathrm{I}}$.

(a) Distributed energy station

Figure 6.16 describes schematically the illustrative diagram of the DES. In DEN i, each DES will import natural gas, $G_{i,k}$, for providing EUs with E&C energy through a gas turbine (GT), a heat recovery steam generator (HRSG), a absorption chiller (AC), and a reciprocating chiller (RC). In addition, electricity, $E_{i,k}$, may also need to be imported from the electricity network, if a more convenient generation way or more profit can be achieved by doing so. The energy generation and conversion in each DES can be expressed in the following matrix form:

$$\begin{bmatrix} e^{\mathrm{e}}_{i,k} \\ e^{\mathrm{c}}_{i,k} \end{bmatrix} = \begin{bmatrix} 1 & (1-\lambda_{i,k})\eta^{\mathrm{e}}_{\mathrm{GT},i} \\ 0 & \lambda_{i,k}\eta^{\mathrm{e}}_{\mathrm{GT},i}\mathrm{COP}_{\mathrm{RC},i} + \eta^{\mathrm{h}}_{\mathrm{GT},i}\eta_{\mathrm{HR},i}\mathrm{COP}_{\mathrm{AC},i} \end{bmatrix} \begin{bmatrix} E_{i,k} \\ G_{i,k} \end{bmatrix} \tag{6.44}$$

where $\eta^{\mathrm{e}}_{\mathrm{GT},i}$ and $\eta^{\mathrm{h}}_{\mathrm{GT},i}$ represent the electrical and thermal efficiencies of GT, respectively. $\eta_{\mathrm{HR},i}$ represents the efficiency of HRSG. $\mathrm{COP}_{\mathrm{RC},i}$ and $\mathrm{COP}_{\mathrm{AC},i}$ are the coefficient of performance (COP) of reciprocating chiller and absorption chiller, respectively. We assume that $\eta^{\mathrm{e}}_{\mathrm{GT},i}$, $\eta^{\mathrm{h}}_{\mathrm{GT},i}$, $\mathrm{COP}_{\mathrm{RC},i}$ and $\mathrm{COP}_{\mathrm{AC},i}$ for all DESs in each DEN are the same. $\lambda_{i,k} \in [0, 1]$ is the scheduling factor of DES k in ith DEN, representing the proportion of electricity generated by GT to be distributed to generate cooling energy.

As one of the profit organization in the IES, each DES tends to achieve its own maximum profit, $U_{i,k}$, by setting the scheduling factor. In addition to the pursuit of profits, all DES $k \in \Bbbk_i$, in DEN, i also need to make sure the local supply and demand balance. Hence

$$\max_{\lambda_{i,k},\, p^{\mathrm{e}}_i,\, p^{\mathrm{c}}_i} U_{i,k} = p^{\mathrm{e}}_i e^{\mathrm{e}}_{i,k} + p^{\mathrm{c}}_i e^{\mathrm{c}}_{i,k} - (c^{\mathrm{e}}_i E_{i,k} + c^{\mathrm{g}}_i G_{i,k} + c^{\mathrm{fix}}_{i,k}) \tag{6.45}$$

$$\text{s.t.} \quad \sum_{k\in\Bbbk_i} e^{\mathrm{e}}_{i,k} = \sum_{n\in\aleph_i} d^{\mathrm{e}}_{i,n} \tag{6.46}$$

$$\sum_{k\in\Bbbk_i} e^{\mathrm{c}}_{i,k} = \sum_{n\in\aleph_i} d^{\mathrm{c}}_{i,n} \tag{6.47}$$

$$p^{\mathrm{e}}_i \ge 0,\ p^{\mathrm{c}}_i \ge 0,\ 0 \le \lambda_{i,k} \le 1,\ \forall k \in \Bbbk_i,\ \forall i \in \tilde{\mathrm{I}} \tag{6.48}$$

for DES k, $\forall k \in \Bbbk_i$, where p^{e}_i and p^{c}_i are the unit prices of E&C energy, respectively. c^{e}_i and c^{g}_i denote the node prices of E&G, respectively, which are the importing energy prices determined by the place where the ith DEN locates, in the utility networks. $c^{\mathrm{fix}}_{i,k}$ is the fixed cost of DES k. $d^{\mathrm{e}}_{i,n}$ and $d^{\mathrm{c}}_{i,n}$ denote the E&C energy demands of EU n, respectively.

(b) Energy users

EUs are nonprofit individuals, i.e., pure energy consumers, in the IES. Each EU aims at obtaining its own maximum welfare, $W_{i,n}$, in the energy trading by deciding on the amount of E&C energy to be consumed. Hence, the optimization problem of EU n, $\forall n \in \aleph_i$ is

$$\begin{aligned}\max_{d^{\mathrm{e}}_{i,n},\, d^{\mathrm{c}}_{i,n}} W_{i,n} &= \left[v^{\mathrm{e}}_{i,n} d^{\mathrm{e}}_{i,n} - \frac{u^{\mathrm{e}}_{i,n}}{2} (d^{\mathrm{e}}_{i,n})^2 \right] \\ &\quad + \left[v^{\mathrm{c}}_{i,n} d^{\mathrm{c}}_{i,n} - \frac{u^{\mathrm{c}}_{i,n}}{2} (d^{\mathrm{c}}_{i,n})^2 \right] - (p^{\mathrm{e}}_i d^{\mathrm{e}}_{i,n} + p^{\mathrm{c}}_i d^{\mathrm{c}}_{i,n})\end{aligned} \tag{6.49}$$

$$\text{s.t.} \quad d^{\mathrm{e}}_{i,n} \ge d^{\mathrm{e,min}}_{i,n},\ d^{\mathrm{c}}_{i,n} \ge 0,\ \forall n \in \aleph_i,\ \forall i \in \tilde{\mathrm{I}} \tag{6.50}$$

where $v^{\mathrm{e}}_{i,n}$, $u^{\mathrm{e}}_{i,n}$, $v^{\mathrm{c}}_{i,n}$ and $u^{\mathrm{c}}_{i,n}$ are the satisfaction parameters of the quadratic utility functions of consuming E&C energy, respectively. $d^{\mathrm{e,min}}_{i,n}$ describes the minimum electricity demand of EU n.

6.3.2 *Energy Trading Game of DESs and EUs*

This section presents a hierarchical Stackelberg game (Tushar et al. 2014) model developed for analyzing the transactions of a variety of energies between DESs and EUs in each DEN. The aforementioned game is a submodular game played by DESs (Topkis 1979). The unit prices of E&C energy is drawn up by DESs, then EUs make a choice about the amount of energies after receiving the price information. Thus, the DESs play the role of forerunners, and the EUs are latecomers. The Stackelberg Equilibrium is obtained in a closed form and its uniqueness is corroborated. In addition, the CDCENs which are related to the amounts and prices of E&G that DESs purchase from utility networks are obtained by the game. The detailed introduction of the hierarchical Stackelberg game will be discussed as follows.

As has been described in Sect. 6.3.1.3, in DEN i, $\forall i \in \tilde{\mathrm{I}}$, DESs trade E&C energy with their local EUs under the environment of perfectly competitive energy market. In order to track the optimal behavior of each participant in the energy trading, we establish a Stackelberg game model for analyzing the problem by respectively endowing DESs and EUs as multiple forerunners and multiple latecomers in each DEN $i \in \tilde{\mathrm{I}}$. The Stackelberg game model can be defined as

$$\begin{aligned}\Psi_i = \{&\aleph_i \cup \Bbbk_i, \{s_{i,n}\}_{n\in\aleph_i}, \{S_{i,k}\}_{k\in\Bbbk_i}, \{W_{i,n}\}_{n\in\aleph_i},\\ &\{U_{i,k}\}_{k\in\Bbbk_i}\},\ \forall i \in \tilde{\mathrm{I}}\end{aligned} \tag{6.51}$$

where $\aleph_i$ and $\Bbbk_i$ are the sets of DESs and EUs in the ith DEN, respectively. $s_{i,n}$ is the strategy set of EU $n \in \aleph_i$. $d^{\mathrm{e}}_{i,n}$ and $d^{\mathrm{c}}_{i,n}$. $S_{i,k}$ is the strategy set of each DES $k \in \Bbbk_i$, including the unit electricity price p^{e}_i, unit cooling price p^{c}_i, and scheduling factor $\lambda_{i,k}$. $W_{i,n}$ and $U_{i,k}$ represent the payoff functions of EU n and DES k as shown in (6.49) and (6.45), respectively.

6.3.2.1 Demand Side Analysis

In the energy trading, EUs individually determine their optimal E&C energy demands in response to the prices set by DESs. It can be seen that (6.49) is strictly concave in both $d^{\mathrm{e}}_{i,n}$ and $d^{\mathrm{c}}_{i,n}$. Hence, given forerunners' prices strategy, the first-order optimality of problem (6.49) gives the efficient solutions of $d^{\mathrm{e}}_{i,n}$ and $d^{\mathrm{c}}_{i,n}$ with respect to p^{e}_i and p^{c}_i, respectively, which are shown as follows:

$$\begin{cases} d^{\mathrm{e}}_{i,n} = \dfrac{v^{\mathrm{e}}_{i,n}}{u^{\mathrm{e}}_{i,n}} - \dfrac{1}{u^{\mathrm{e}}_{i,n}} p^{\mathrm{e}}_i,\ \forall n \in \aleph_i,\ \forall i \in \tilde{\mathrm{I}} \\ d^{\mathrm{c}}_{i,n} = \dfrac{v^{\mathrm{c}}_{i,n}}{u^{\mathrm{c}}_{i,n}} - \dfrac{1}{u^{\mathrm{c}}_{i,n}} p^{\mathrm{c}}_i,\ \forall n \in \aleph_i,\ \forall i \in \tilde{\mathrm{I}} \end{cases} \tag{6.52}$$

The inequality constraints as follows are the achievement of (6.52) in (6.50) and employed to meet the feasible region constraints on the consumption of E&C energy of each EU.

$$\begin{aligned} p_i^{\mathrm{e}} &\le \min(v_{i,n}^{\mathrm{e}} - u_{i,n}^{\mathrm{e}} d_{i,n}^{\mathrm{e,min}}) \ , \forall n \in \aleph_i \ , \forall i \in \tilde{\mathrm{I}} \\ p_i^{\mathrm{c}} &\le \min(v_{i,n}^{\mathrm{c}}) \qquad\qquad\quad , \forall n \in \aleph_i \ , \forall i \in \tilde{\mathrm{I}} \end{aligned} \tag{6.53}$$

6.3.2.2 Supply Side Analysis

For each DES $k \in \Bbbk_i$, the constrained optimization problem (6.45) can be transformed into a non-constrained one as followed by broadening the feasible region constraints on p_i^{e}, p_i^{c} and $\lambda_{i,k}$.

$$\begin{aligned} L_{i,k} &= p_i^{\mathrm{e}} e_{i,k}^{\mathrm{e}} + p_i^{\mathrm{c}} e_{i,k}^{\mathrm{c}} - (c_i^{\mathrm{e}} E_{i,k} + c_i^{\mathrm{g}} G_{i,k} + c_{i,k}^{\mathrm{fix}}) \\ &\quad + \mu_{k,1}^i \left(\sum_{k \in \Bbbk_i} e_{i,k}^{\mathrm{e}} - \sum_{n \in \aleph_i} d_{i,n}^{\mathrm{e}} \right) + \mu_{k,2}^i \left(\sum_{k \in \Bbbk_i} e_{i,k}^{\mathrm{c}} - \sum_{n \in \aleph_i} d_{i,n}^{\mathrm{c}} \right) \end{aligned} \tag{6.54}$$

where $\mu_{k,1}^i$ and $\mu_{k,2}^i$, $\forall k \in \Bbbk_i$, are the Lagrange multipliers. The complementarity slackness conditions are

$$\begin{aligned} &\mu_{k,1}^i \left(\sum_{k \in \Bbbk_i} e_{i,k}^{\mathrm{e}} - \sum_{n \in \aleph_i} d_{i,n}^{\mathrm{e}} \right) = 0, \ \sum_{k \in \Bbbk_i} e_{i,k}^{\mathrm{e}} - \sum_{n \in \aleph_i} d_{i,n}^{\mathrm{e}} = 0 \\ &\mu_{k,2}^i \left(\sum_{k \in \Bbbk_i} e_{i,k}^{\mathrm{c}} - \sum_{n \in \aleph_i} d_{i,n}^{\mathrm{c}} \right) = 0, \ \sum_{k \in \Bbbk_i} e_{i,k}^{\mathrm{c}} - \sum_{n \in \aleph_i} d_{i,n}^{\mathrm{c}} = 0 \end{aligned} \tag{6.55}$$

Then, by using (6.44) and (6.52) in (6.54), the first-order optimal condition for maximizing $L_{i,k}$ gives

$$\frac{\partial L_{i,k}}{\partial p_i^{\mathrm{e}}} = \eta_{\mathrm{GT},i}^{\mathrm{e}} (1 - \lambda_{i,k}) G_{i,k} + E_{i,k} + \mu_{k,1}^i A_i^{\mathrm{e}} = 0 \tag{6.56}$$

$$\frac{\partial L_{i,k}}{\partial p_i^{\mathrm{c}}} = (D_i \lambda_{i,k} + O_i) G_{i,k} + \mu_{k,2}^i A_i^{\mathrm{c}} = 0 \tag{6.57}$$

$$\frac{\partial L_{i,k}}{\partial \lambda_{i,k}} = (-\eta_{\mathrm{GT},i}^{\mathrm{e}} p_i^{\mathrm{e}} + D_i p_i^{\mathrm{c}} - \mu_{k,1}^i \eta_{\mathrm{GT},i}^{\mathrm{e}} + \mu_{k,2}^i D_i) G_{i,k} = 0 \tag{6.58}$$

$$\frac{\partial L_{i,k}}{\partial \mu_{k,1}^i} = \sum_{k \in \Bbbk_i} \eta_{\mathrm{GT},i}^{\mathrm{e}} (1 - \lambda_{i,k}) G_{i,k} + \sum_{k \in \Bbbk_i} E_{i,k} - B_i^{\mathrm{e}} + A_i^{\mathrm{e}} p_i^{\mathrm{e}} = 0 \tag{6.59}$$

$$\frac{\partial L_{i,k}}{\partial \mu_{k,2}^i} = \sum_{k \in \Bbbk_i} (D_i \lambda_{i,k} + O_i) G_{i,k} - B_i^{\mathrm{c}} + A_i^{\mathrm{c}} p_i^{\mathrm{c}} = 0 \tag{6.60}$$

where $A_i^e=\sum_{n\in\aleph_i}(1/u_{i,n}^e)$, $B_i^e=\sum_{n\in\aleph_i}(v_{i,n}^e/u_{i,n}^e)$, $A_i^c=\sum_{n\in\aleph_i}(1/u_{i,n}^c)$, $D_i=\eta_{GT,i}^e$ $COP_{RC,i}$, $B_i^c=\sum_{n\in\aleph_i}(v_{i,n}^c/u_{i,n}^c)$ and $O_i=\eta_{GT,i}^h\eta_{HR,i}COP_{AC,i}$.

Let $F_i=(\eta_{GT,i}^e)^2/A_i^e$, $Z_i=D_i^2/A_i^c$, $H_i=D_iO_i/A_i^c$ and $R_i=D_iB_i^c/A_i^c-\eta_{GT,i}^e B_i^e/A_i^e$. From (6.56), (6.57), (6.59) and (6.60), we obtain the closed-form expressions of p_i^e, p_i^c and $\lambda_{i,k}$ as follows:

$$\begin{cases}\lambda_{i,k}=\dfrac{A_i^c(K_i+1)(F_i-H_i)G_{i,k}+\eta_{GT,i}^e(K_i+1)E_{i,k}+A_i^eR_i}{A_i^e(K_i+1)(F_i+Z_i)G_{i,k}},\\ p_i^e=-X_{i1}\sum\limits_{k\in\Bbbk_i}G_{i,k}-X_{i2}\sum\limits_{k\in\Bbbk_i}E_{i,k}+X_{i3},\\ p_i^c=-Y_{i1}\sum\limits_{k\in\Bbbk_i}G_{i,k}-Y_{i2}\sum\limits_{k\in\Bbbk_i}E_{i,k}+Y_{i3},\end{cases}\tag{6.61}$$

where $\forall k\in\Bbbk_i$, $\forall i\in\tilde{\mathrm{I}}$ and

$$\begin{aligned}X_{i1}&=\frac{\eta_{GT,i}^e(H_i+Z_i)}{A_i^e(F_i+Z_i)}\\ Y_{i1}&=\frac{O_i(F_i+Z_i)+D_i(F_i-H_i)}{A_i^c(F_i+Z_i)}\\ X_{i2}&=\frac{A_i^e(F_i+Z_i)-(\eta_{GT,i}^e)^2}{(A_i^e)^2(F_i+Z_i)}\\ Y_{i2}&=\frac{D_i\eta_{GT,i}^e}{A_i^eA_i^c(F_i+Z_i)}\\ X_{i3}&=\frac{B_i^e}{A_i^e}+\frac{\eta_{GT,i}^eK_iR_i}{A_i^e(K_i+1)(F_i+Z_i)}\\ Y_{i3}&=\frac{B_i^c}{A_i^c}-\frac{D_iK_iR_i}{A_i^c(K_i+1)(F_i+Z_i)}\end{aligned}$$

Furthermore, here are $G_{i,k}>0$ and $E_{i,k}\geq 0$ to guarantee $\lambda_{i,k}\in[0,1]$, $G_{i,k}\neq 0$.

In the following text, we will focus on analyzing the uniqueness of the Stackelberg Equilibrium of the energy trading game. Moreover, the requirements for validating the optimal strategy of each DES $k\in\Bbbk_i$ as shown in (6.61) will also be derived through analyzing the property of the forerunner's game, which clearly reveals how the CDCENs are obtained.

A. Uniqueness of the Stackelberg Equilibrium

All DESs will charge their local EUs the uniform unit energy prices in perfectly competitive energy markets, i.e., $p_i^e=p_{i,k}^e$ and $p_i^c=p_{i,k}^c$, $\forall k\in\Bbbk_i$, $\forall i\in\tilde{\mathrm{I}}$. Under such market environment, it is evident that the market prices of E&C energy are collectively set by all DESs in the energy trading. Therefore, there exists a price competition that DESs in each DEN needs to compete with everyone else to seek its own individual benefit, thus the price setting game of the DESs is mostly non-cooperative. From analyzing the property of (6.45), it is interesting to find that by strictly monotonic transforming the forerunner's payoff function (Amir et al. 1999;

Milgrom and Shannon 1994), i.e., logarithmic transformation, it satisfies the characteristics of submodular. In microeconomics, submodularity indicates that even any minor increase in one player always leads to the decrease in other marginal benefits (Topkis 1979, 1978).

First, in order to mathematically guarantee a harmony antilogarithm, we take the monotonic transformation of (6.45)

$$\log U_{i,k}^{'} = \log\left(a_{i,k}U_{i,k} + b_{i,k}\right) \tag{6.62}$$

where $a_{i,k} > 0$ and $b_{i,k}$, $\forall i \in \tilde{\mathrm{I}}$, are constants.

Then, by taking the cross-partial derivatives of $\log U_{i,k}^{'}$ with respect to the strategy of DES k, we have

$$\frac{\partial^2 \log U_{i,k}^{'}}{\partial p_i^{\mathrm{e}} \partial p_i^{\mathrm{c}}} = \frac{-a_{i,k}^2 G_{i,k}\left[\eta_{\mathrm{GT},i}^{\mathrm{e}}(1-\lambda_{i,k})G_{i,k} + E_{i,k}\right](D_i\lambda_{i,k} + O_i)}{U_{i,k}^{'2}\ln 10} \tag{6.63}$$

$$\frac{\partial^2 \log U_{i,k}^{'}}{\partial \lambda_{i,k} \partial p_i^{\mathrm{e}}} = \frac{-a_{i,k}G_{i,k}\left[a_{i,k}\eta_{\mathrm{GT},i}^{\mathrm{e}}(D_i + O_i)G_{i,k} + a_{i,k}D_i E_{i,k}\right](p_i^{\mathrm{c}} - \varphi_{i,k})}{U_{i,k}^{'2}\ln 10} \tag{6.64}$$

$$\frac{\partial^2 \log U_{i,k}^{'}}{\partial \lambda_{i,k} \partial p_i^{\mathrm{c}}} = \frac{a_{i,k}G_{i,k}\left[a_{i,k}\eta_{\mathrm{GT},i}^{\mathrm{e}}(D_i + O_i)G_{i,k} + a_{i,k}D_i E_{i,k}\right](p_i^{\mathrm{e}} - \phi_{i,k})}{U_{i,k}^{'2}\ln 10} \tag{6.65}$$

According to the definition of a log-submodular function as described in Topkis (1979), only when (6.63), (6.64) and (6.65) are all non-harmony, $U_{i,k}^{'}$ is log-submodular and therefore, the payoff function of DES k, $U_{i,k}^{'}$, is submodular. It is evident that (6.63) is always non-harmony. In addition, as the denominators of (6.64) and (6.65) are always harmony, their numerators should not be larger than zero. In order to make the numerators of (6.64) and (6.65) less equal to zero, the following inequality constraints must hold:

$$p_i^{\mathrm{e}} \leq \min\left(\phi_{i,k}\right),\ \forall k \in \mathbb{k}_i,\ \forall i \in \tilde{\mathrm{I}} \tag{6.66}$$

$$p_i^{\mathrm{c}} \geq \max\left(\varphi_{i,k}\right),\ \forall k \in \mathbb{k}_i,\ \forall i \in \tilde{\mathrm{I}} \tag{6.67}$$

where

$$\phi_{i,k} = \frac{a_{i,k}D_i(c_i^{\mathrm{e}}E_{i,k} + c_i^{\mathrm{g}}G_{i,k} + c_{i,k}^{\mathrm{fix}}) - b_{i,k}D_i}{a_{i,k}\eta_{\mathrm{GT},i}^{\mathrm{e}}(D_i + O_i)G_{i,k} + a_{i,k}D_i E_{i,k}}$$

$$\varphi_{i,k} = \frac{a_{i,k}\eta_{\mathrm{GT},i}^{\mathrm{e}}(c_i^{\mathrm{e}}E_{i,k} + c_i^{\mathrm{g}}G_{i,k} + c_{i,k}^{\mathrm{fix}}) - b_{i,k}\eta_{\mathrm{GT},i}^{\mathrm{e}}}{a_{i,k}\eta_{\mathrm{GT},i}^{\mathrm{e}}(D_i + O_i)G_{i,k} + a_{i,k}D_i E_{i,k}}$$

If and only if both (6.66) and (6.67) are hold, $U_{i,k}$ is submodular. The requirements of a function to be submodular can be found in Topkis (1979).

Equations (6.66) and (6.67) limit the maximum electrical energy price and the minimum cooling energy price. In addition, we should also meet (6.48) and (6.53).

Collectively meeting the abovementioned constraints, there are

$$0 \leq p_i^e \leq \psi_{i,k} \quad , \forall n \in \aleph_i, \ \forall k \in \Bbbk_i, \ \forall i \in \tilde{\mathrm{I}} \tag{6.68}$$

$$\max(\varphi_{i,k}) \leq p_i^c \leq \min(v_{i,n}^c), \ \forall n \in \aleph_i, \ \forall k \in \Bbbk_i, \ \forall i \in \tilde{\mathrm{I}} \tag{6.69}$$

where $\psi_{i,k} = \min\left(\min(v_{i,n}^e - u_{i,n}^e d_{i,n}^{e,\min}), \ \min(\phi_{i,k})\right)$.

Since all the feasible regions of p_i^e and p_i^c, both of which are non-empty, convex, and compact, have been obtained, the submodularity of the DESs' price setting game can be ensured according to Theorem 1 in Wei et al. (2017).

Submodularity of the forerunner's game ensures the existence of pure strategic Nash Equilibriums (Topkis 1979). Since the latecomers always react in a unique response to the forerunners' actions, the Stackelberg Equilibrium of the energy trading game between DESs and EUs in each DEN $i \in \tilde{\mathrm{I}}$ is unique according to Theorem 2 in Wei et al. (2017). The detailed proof of the submodularity of DESs' price setting game and the uniqueness of the Stackelberg Equilibrium of the energy trading game can be referred to Wei et al. (2017).

In the course of the game, first, the initial values of p_i^e, p_i^c are randomly set in the feasible region, as (6.68) and (6.69) shown, and the E&C energy demands $d_{i,n}^e$ and $d_{i,n}^c$ of EUs can be obtained through the trade between DES and EUs. Second, p_i^e and p_i^c are updated by the amount of unbalance between supply and demand and p_{i+1}^e and p_{i+1}^c are obtained in the new iteration. Then the second step will be repeated, until, the convergence conditions $|p_{i+1}^e - p_i^e| < \theta$ and $|p_{i+1}^c - p_i^c| < \theta$ are simultaneously met and finally, p_i^e and p_i^c are determined.

B. Coupling demand constraints of electricity and natural gas

As mentioned before, the Stackelberg Equilibrium of the hierarchical Stackelberg game model developed for analyzing the energy trading problem between DESs and EUs in each district energy network DEN includes three parameters: scheduling factor $\lambda_{i,k}$, unit price of electrical energy p_i^e, and unit price of cooling energy p_i^c. And, the equilibrium solution has a direct effect on the amount of E&G produced by the E&G networks.

In terms of the scheduling factor, it can be seen from the expression of $\lambda_{i,k}$ as shown in (6.61) that there are only two exogenous parameters, i.e., $G_{i,k}$ and $E_{i,k}$, while all the others are constant parameters of DESs and EUs. Thus, the requirements for $\lambda_{i,k}$ are set to be within [0, 1] as for $G_{i,k}$ and $E_{i,k}$, which are given as follows:

$$\begin{cases} A_i^e(H_i + Z_i)G_{i,k} - \eta_{\mathrm{GT},i}^e E_{i,k} \geq \dfrac{A_i^e R_i}{K_i + 1}, \ \forall k \in \Bbbk_i, \ \forall i \in \tilde{\mathrm{I}} \\ A_i^e(H_i - F_i)G_{i,k} - \eta_{\mathrm{GT},i}^e E_{i,k} \leq \dfrac{A_i^e R_i}{K_i + 1}, \ \forall k \in \Bbbk_i, \ \forall i \in \tilde{\mathrm{I}} \end{cases} \tag{6.70}$$

Similarly, close observation of the closed-form expressions of E&C energy prices reveals that in each DEN $i \in \tilde{\mathrm{I}}$, the energy prices are affected by all the input amounts of natural gas and electricity of all the local DESs, i.e., $G_{i,k}$ and $E_{i,k}$, $\forall k \in \Bbbk_i$.

Equations (6.68) and (6.69) indicate the feasible regions of p_i^{e} and p_i^{c}, respectively, which are deduced from the feasible region constraints on the E&C energy of each EU and the requirements to ensure a submodular $U_{i,k}$. Equations (6.68) and (6.69) reflect how the amounts and prices of E&G in wholesale market influence retail prices of E&C energy traded between DESs and EUs. By substituting the closed-form expressions of p_i^{e} and p_i^{c} into (6.68) and (6.69), respectively, we have

$$X_{i3}-\psi_{i,k} \leq X_{i1}\sum_{k\in\Bbbk_i} G_{i,k}+X_{i2}\sum_{k\in\Bbbk_i} E_{i,k} \leq X_{i3} \tag{6.71}$$

$$Y_{i3}-\min\left(v_{i,n}^{\mathrm{c}}\right) \leq Y_{i1}\sum_{k\in\Bbbk_i} G_{i,k}+Y_{i2}\sum_{k\in\Bbbk_i} E_{i,k} \leq Y_{i3}-\max\left(\varphi_{i,k}\right) \tag{6.72}$$

Therefore, the equations in CDCENs are simultaneous solutions of each parameter's closed-form expression and its feasible region constraints. Equation (6.70) is the new constraints about the amount of E&G according to scheduling factor $\lambda_{i,k}$, (6.71) is the new constraint about the amount of E&G according to unit price of electrical energy p_i^{e}, and (6.72) is the new constraint about the amount of E&G according to unit price of cooling energy p_i^{c}. Equations (6.71) and (6.72) jointly restrict the total amounts of natural gas and electricity purchased by DESs in response to the prices in each DEN, which indicates the correlation between the wholesale prices and DESs' demands of natural gas and electricity. Equation (6.70) indicates that unless the input amounts of natural gas and electricity of each DES $k \in \Bbbk_i$ supplied by utility networks satisfy these two ranges, the optimal strategy of $\lambda_{i,k}$ to be within [0, 1] will be invalid. Equations (6.70), (6.71), and (6.72) are therefore the CDCENs, which are obtained based on the requirements for the validations of each participant's strategy and the requirements for satisfying the submodularity of the game. It is shown that the CDCENs are composed of the parameters of both the EUs' demand characteristics and DESs' operating characteristics. By reflecting the relationship between the amount and prices of E&G traded between DESs and utility networks, the CDCENs depict the demand characteristics of DESs and denote the interactions between DESs and utility networks.

6.3.3 The Hybrid Multi-objective Optimization and Game-Theoretic Approach

A hybrid multi-objective optimization and game-theoretic approach (HMOGTA) is presented to solve a coordinated operation strategy of the IES. For the conflicting benefits of E&G networks considering the CDCENs, the multi-objective group search optimizer with adaptive covariance and chaotic search (MGSOACC) (Wei et al. 2016) is applied to generate efficient solutions and simultaneously optimize their own operating costs. Finally, the evidential reasoning (ER) approach (Yang and Xu 2002) is employed to select a final operation solution. Figure 6.17 illustrates the flowchart of HMOGTA and the detailed introduction of the HMOGTA will be shown as follows.

6.3.3.1 Optimization of the Conflicting Costs of Electricity and Gas Networks

As both the E&G can satisfy the demands of DESs, there exists a substitutive relationship between the amount of them. The CDCENs demonstrating this kind of substitutive relationship reflect the conflicting relationship between the operating costs of E&G networks. Therefore, we develop a multi-objective optimization model for optimizing the conflicting costs of E&G networks:

$$\min\ [J_{\mathrm{g}}(U_{\mathrm{g}}, X_{\mathrm{g}}, W_{\mathrm{g}}), J_{\mathrm{e}}(U_{\mathrm{e}}, X_{\mathrm{e}}, W_{\mathrm{e}})] \tag{6.73}$$

$$\begin{aligned} \text{s.t. } & g_{\mathrm{g}}(U_{\mathrm{g}}, X_{\mathrm{g}}, W_{\mathrm{g}}) = 0 \\ & h_{\mathrm{g}}(U_{\mathrm{g}}, X_{\mathrm{g}}, W_{\mathrm{g}}) \le 0 \\ & g_{\mathrm{e}}(U_{\mathrm{e}}, X_{\mathrm{e}}, W_{\mathrm{e}}) = 0 \\ & h_{\mathrm{e}}(U_{\mathrm{e}}, X_{\mathrm{e}}, W_{\mathrm{e}}) \le 0 \\ & [W_{\mathrm{g}}, W_{\mathrm{e}}] \in \text{CDCENs} \end{aligned} \tag{6.74}$$

where W_{g} denotes the vector of the interactive variables between natural gas network and DESs. W_{e} denotes the vector of the interactive variables between electricity network and DESs, i.e., $W_{\mathrm{g}}^{\mathrm{T}} = [G_{1,1} \cdots G_{I,K_I}, c_1^{\mathrm{g}} \cdots c_I^{\mathrm{g}}]$ and $W_{\mathrm{e}}^{\mathrm{T}} = [E_{1,1} \cdots E_{I,K_I}, c_1^{\mathrm{e}} \cdots c_I^{\mathrm{e}}]$. In this section, locational marginal prices (LMPs) are employed as wholesale prices of E&G, the detailed introduction of which can be referred to Dommel and Tinney (1968). As have been shown in (6.70), (6.71), and (6.72), the CDCENs, with respect to W_{g} and W_{e}, demonstrate the conflicting relationship between the operating costs of E&G networks. Apart from g_{g}, h_{g}, g_{e}, and h_{e}, the optimal operation of E&G networks should also consider the demand characteristics of DESs, i.e., the CDCENs. The CDCENs closely interconnect the utility companies and DESs in terms of their interactions and can fairly help optimize individual objective of each participant in the IES.

In this section, the MGSOACC and the ER approach are employed to deal with the minimization of the conflicting costs of E&G networks. The MGSOACC, a multi-objective optimization algorithm developed based on group searching behaviors of animals, adopts adaptive covariance and chaotic search as searching strategies of scroungers and rangers, respectively. The MGSOACC is employed in this chapter for its better performance in terms of its convergence (Wei et al. 2016). Moreover, the ER approach proposed in Yang and Xu (2002) is utilized for multi-attribute decision-making with uncertainty. Multi-objective optimization and decision-making methods have been widely used to solve the optimal scheduling problem (Zheng et al. 2017; Li et al. 2014; Zheng et al. 2015). In this section, the MGSOACC is used to simultaneously optimize the operating costs of E&G networks in order to get Pareto-optimal

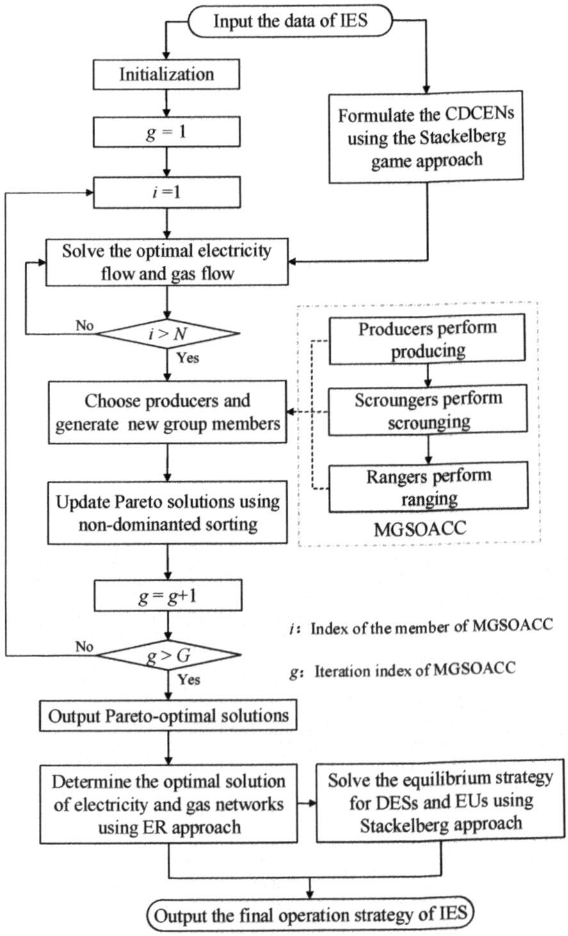

Fig. 6.17 The flowchart of hybrid multi-objective optimization and game-theoretic approach

frontier. Afterward, the ER approach is employed to select a final solution from Pareto-optimal frontier. The detailed introduction of the MGSOACC and the ER approach can be referred to Wei et al. (2016) and Yang and Xu (2002), respectively.

6.3.3.2 Solution Method of the HMOGTA

A hybrid multi-objective optimization and game-theoretic approach (HMOGTA) is presented to figure out the optimal operation problem of the IES. The energy trading problem between DESs and EUs in each DEN is solved by the Stackelberg game approach with the CDCENs obtained. The CDCENs, with regards to the interactive variables between utility networks and DESs, can reveal the conflicting relationship between the operating costs of E&G networks. Afterward, the minimization of

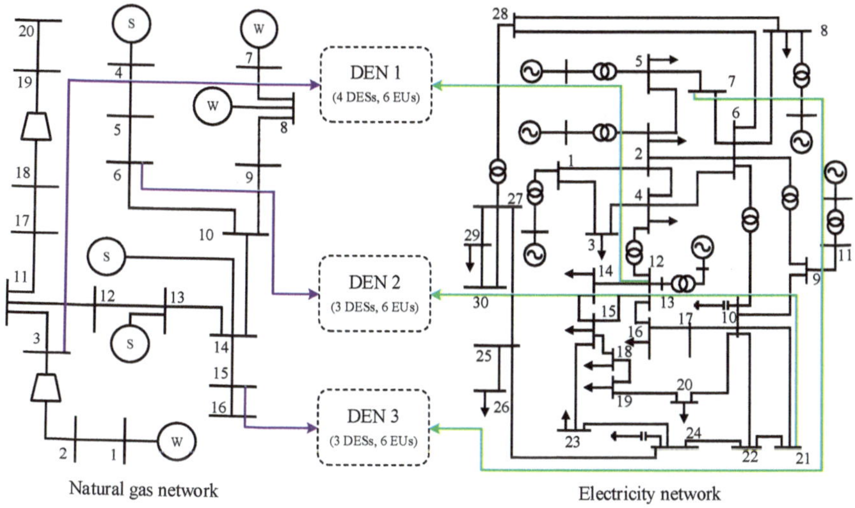

Fig. 6.18 Integrated energy system composed of a 20-node natural gas network, a 30-bus electricity network, and three DENs

conflicting costs of E&G networks in consideration of the CDCENs is solved by the MGSOACC and the ER approach.

6.3.4 Simulation Studies

6.3.4.1 System Description

In this section, simulation studies are conducted on a test IES consisting of natural gas network, electricity network, and DENs. As has been shown in Fig. 6.18, the test IES consists of a 20-node natural gas network, a modified IEEE 30-bus system, and 3 DENs. It is worth mentioning that in the test IES, the coupling between electricity and natural gas network are not considered. Electricity and natural gas network individually transmit energy to each DEN, and all the units in electricity network are assumed to be thermal power units.

It can be seen from the figure that 3 DENs connected to the electricity network at buses 7, 12, and 21 as electricity loads, while they are connected to the natural gas network at nodes 3, 6, and 15 serving as gas loads. As has been displayed in Fig. 6.18, 6 EUs play the role of participants in energy trading in local energy market in each DEN. In addition, 4 DESs are set to compete with each other in DEN 1 while 3 DESs are considered in each of the other two DENs. Therefore, there are totally 10 DESs who not only serve as the loads in both the E&G networks but also perform as distributed energy suppliers for their local EUs.

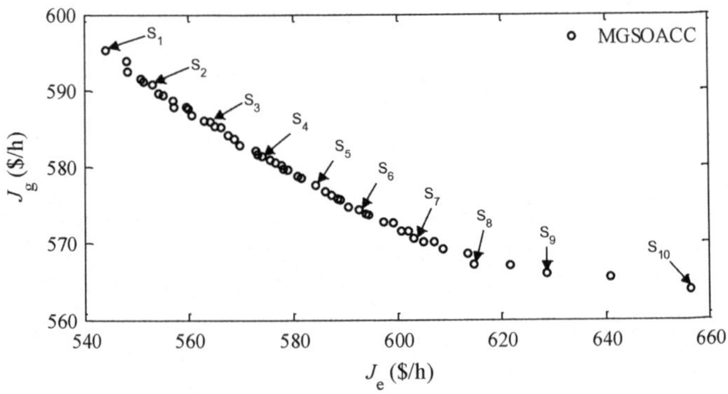

Fig. 6.19 Efficient solutions obtained by the MGSOACC of the IES

6.3.4.2 Results Analysis

The hybrid multi-objective optimization and Stackelberg game approach proposed aforementioned is utilized to obtain the operation solution of the IES for fairly treating each individual participant. The number of function evaluations of the multi-objective optimization algorithm, MGSOACC, is set as 20,000. The optimization results of the operating costs of E&G networks are shown in Fig. 6.19.

Figure 6.19 illustrates the conflicting relationship of minimizing the operating costs of the E&G networks, with consideration of the CDCENs.

Such a conflicting relationship reflects the fact that simultaneously minimizing the operating costs of the E&G networks is impossible. Therefore, the ISO needs to determine a final operation solution, which ensures the fair treatment to the E&G networks. To achieve this, the ER approach is utilized to make the final decision. To clearly present the ER, 10 efficient solutions obtained by the MGSOACC are selected, namely $[S_1, S_2, \ldots, S_{10}]$. As the economic benefit is the key index to be considered in this study, the operating cost of each network is set as the basic attribute in the ER approach and the relative weights are equally set as $\omega = [0.5, 0.5]$. In addition, the evaluation grades are set as $H = \{\text{Poor, Unsatisfied, Normal, Good, Excellent}\}$.

By applying the procedure of ER, the ranking of the 10 selected alternatives ranked based on their averaged utilities can be obtained, which is summarized in Table 6.6. It can be seen from the table that, solution S_5 obtains the highest averaged utility, which is therefore determined as the final operation solution for the E&G networks. Table 6.7, Figs. 6.20 and 6.21 display the results of DESs and EUs in each DEN as for solution S_5. We can see that even though the energy demands of EUs in DEN 1 are obviously larger than those in DEN 2 and DEN 3, the profits of DESs in DEN 1 are lower than those in the other two DENs. That is because the number of DESs in DEN 1 is larger than that in both DEN 2 and DEN 3, which makes the competition between DESs in DEN 1 intenser. Additionally, it can be observed that the profits of all DESs in DEN 3 are lower than those of DESs in DEN 2. As is shown in Table 6.7, the retail prices of E&C energy in DEN 3 are nearly the same as those in DEN 2,

Table 6.6 The ranking of the 10 selected efficient solutions

Solution	S_1	S_2	S_3	S_4	S_5	S_6	S_7	S_8	S_9	S_{10}
J_e	544.120	553.117	562.993	574.003	584.350	597.339	604.987	613.565	628.581	656.324
J_g	595.355	590.785	586.020	581.406	577.662	572.718	570.135	568.623	565.952	563.898
μ_{max}	0.567	0.589	0.613	0.630	0.640	0.626	0.606	0.600	0.594	0.544
μ_{min}	0.471	0.494	0.520	0.545	0.555	0.537	0.521	0.510	0.455	0.448
μ_{avg}	0.519	0.541	0.566	0.588	0.598	0.581	0.564	0.555	0.524	0.486
Ranking	9	7	4	2	1	3	5	6	8	10

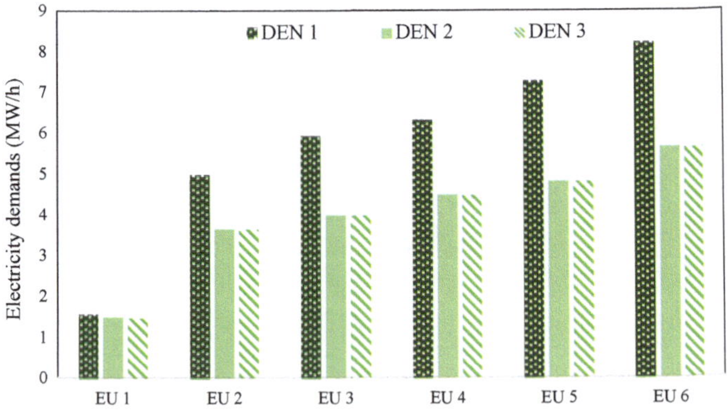

Fig. 6.20 Electricity demands as for S_5

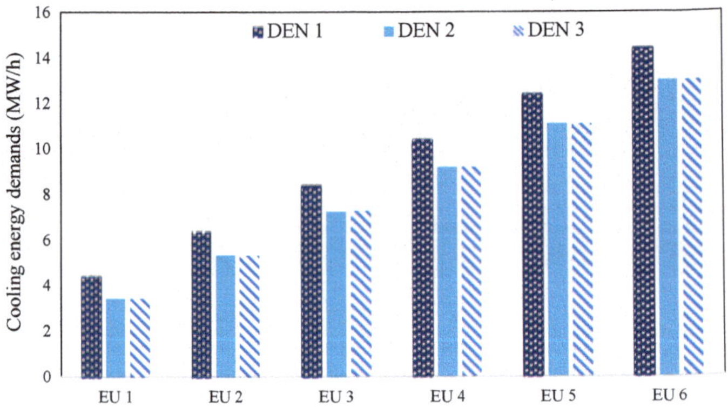

Fig. 6.21 Cooling energy demands as for S_5

respectively, which means that the energy demands of EUs and total incomes of DESs in these two areas are almost the same. Therefore, it is the cost differences that contribute to the profit differences of DESs in these two areas. Since EUs' cooling demands, which are all satisfied by consuming natural gas, are larger than electricity demands, the amount of natural gas purchased by DESs is larger than the amount of electricity. Thus the costs of DESs in DEN 3 are higher than those of DESs in DEN 2 for the higher wholesale price of natural gas in DEN 3.

In order to demonstrate the conflicting relationship of the operating costs between the E&G networks subject to the CDCENs, and the necessity of employing an ISO to coordinate their benefits, we compare the results obtained by the proposed HMOGTA with those of the following cases:

S_e: Minimizing the operating cost of electricity network (OCEN).
S_g: Minimizing the operating cost of natural gas network (OCNN).
S_{eg}: Minimizing the aggregate operating costs of E&G networks (AOCENN).

Table 6.7 The results of each DES as for S_5

Item	DEN 1				DEN 2			DEN 3		
	DES 1	DES 2	DES 3	DES 4	DES 1	DES 2	DES 3	DES 1	DES 2	DES 3
$U_{i,k}$ ($/h$)	12.783	13.108	13.215	12.508	15.348	14.961	14.814	14.182	14.572	14.413
$E_{i,k}$ (MW)	4.627	4.012	6.887	6.250	4.681	8.731	2.089	0.669	9.409	5.830
$G_{i,k}$ (MW)	11.266	13.824	14.907	11.338	15.088	14.297	11.200	12.592	12.094	14.870
p_i^e ($/MW)	3.498				3.390			3.397		
p_i^c ($/MW)	1.795				1.801			1.803		
c_i^e ($/MW)	3.705				3.706			3.591		
c_i^g ($/MW)	1.639				1.621			1.721		

Table 6.8 Comparison of operating costs of E&G networks

Item	Notation	J_e ($/h$)	J_g ($/h$)
HMOGTA	S_5	584.35	577.66
Minimizing OCEN	S_e	386.35	665.68
Minimizing OCNN	S_g	760.99	534.04
Minimizing AOCENN	S_{eg}	386.35	654.32

Table 6.9 The profits of DESs as for S_5, S_e, S_g, and S_{eg}

Item	DEN 1				DEN 2			DEN 3		
	$U_{1,1}$	$U_{1,2}$	$U_{1,3}$	$U_{1,4}$	$U_{2,1}$	$U_{2,2}$	$U_{2,3}$	$U_{3,1}$	$U_{3,2}$	$U_{3,3}$
S_5	12.78	13.11	13.22	12.51	15.44	14.96	14.82	14.18	14.57	14.41
S_e	10.14	9.97	10.09	10.16	3.61	8.90	4.47	3.99	6.36	5.18
S_g	9.98	9.86	7.42	5.91	9.68	12.93	11.41	13.75	11.05	10.25
S_{eg}	11.96	11.84	11.19	11.53	14.10	13.87	13.89	11.31	11.04	10.93

Table 6.10 The amount of electricity traded between electricity network and DESs

Item	DEN 1				DEN 2			DEN 3		
	$E_{1,1}$	$E_{1,2}$	$E_{1,3}$	$E_{1,4}$	$E_{2,1}$	$E_{2,2}$	$E_{2,3}$	$E_{3,1}$	$E_{3,2}$	$E_{3,3}$
S_5	4.627	4.012	6.887	6.250	4.681	8.731	2.089	0.669	9.409	5.830
S_e	0.000	0.000	0.000	0.000	0.000	0.000	0.000	0.000	0.000	0.000
S_g	5.987	6.005	9.810	12.086	10.571	5.742	7.795	4.548	9.158	10.401
S_{eg}	0.000	0.000	0.000	0.000	0.000	0.000	0.000	0.000	0.000	0.000

Table 6.11 The amount of natural gas traded between natural gas network and DESs

Item	DEN 1				DEN 2			DEN 3		
	$G_{1,1}$	$G_{1,2}$	$G_{1,3}$	$G_{1,4}$	$G_{2,1}$	$G_{2,2}$	$G_{2,3}$	$G_{3,1}$	$G_{3,2}$	$G_{3,3}$
S_5	11.266	13.824	14.907	11.338	15.088	14.297	11.200	12.592	12.094	14.870
S_e	27.969	30.559	22.036	15.352	34.922	15.801	31.158	29.936	22.117	25.528
S_g	6.186	6.188	6.574	6.805	7.844	7.276	7.516	7.140	7.674	7.824
S_{eg}	28.475	27.888	13.916	25.102	25.924	20.715	25.528	22.808	24.662	24.688

Table 6.12 The wholesale prices and the retail prices of energies

Item	DEN 1				DEN 2				DEN 3			
	p_1^e	p_1^c	c_1^e	c_1^g	p_2^e	p_2^c	c_2^e	c_2^g	p_3^e	p_3^c	c_3^e	c_3^g
S_e	3.504	1.797	3.497	1.836	3.027	1.721	3.528	1.878	3.192	1.758	3.483	1.993
S_g	3.525	1.801	4.409	1.314	3.400	1.804	4.411	1.237	3.400	1.804	4.267	1.345
S_{eg}	3.525	1.801	3.497	1.781	3.400	1.804	3.528	1.753	3.400	1.804	3.483	1.872

Note that the CDCENs are taken into consideration in all the four cases and all parameters remain the same. It can be seen from S_e and S_g, as shown in Table 6.8, that the solutions obtained by single-objective optimization methods achieve the lowest operating cost of the corresponding utility network at the price of largely increasing the cost of the other one, respectively (Table 6.9). Tables 6.10 and 6.11 display the amount of E&G traded between utility networks and DESs in different cases, respectively, which demonstrate the fact that in order to diminish the operating cost as much as possible, the operator of either the electricity or the natural gas network will be prone to reduce the amount of energy traded without considering the operation strategy of the other one. It is no doubt that reducing the load reduces the operating cost of the corresponding utility network. As a result, as long as the CDCENs are satisfied, any single-objective optimization method considering merely the minimum operating cost of a specified utility network will increase the burden of the other one by forcing it to supply more energy regardless of how high the operating cost it may suffer. However, for any rational participants in the IES, such kind of bias is unacceptable as the single-objective optimization method fails to ensure a fair treatment, which may eventually incur severe envy between them.

In terms of S_{eg}, when optimizing the aggregate operating costs of E&G networks, only natural gas will be traded with DESs as natural gas generation is much cheaper than electricity. Therefore, unless the natural gas network is unable to supply enough energy for DESs, the minimum aggregate operating costs of E&G networks will be optimized merely by trading natural gas with DESs. Note that differing from S_e, S_{eg} also ensures the optimal gas flow, by which the lower operating cost of natural gas network under the condition of solely satisfying the CDCENs can be achieved. But the operating cost of the natural gas network is still much higher than that obtained by S_5. In terms of this, the solution S_5 obtained by the proposed HMOGTA successfully

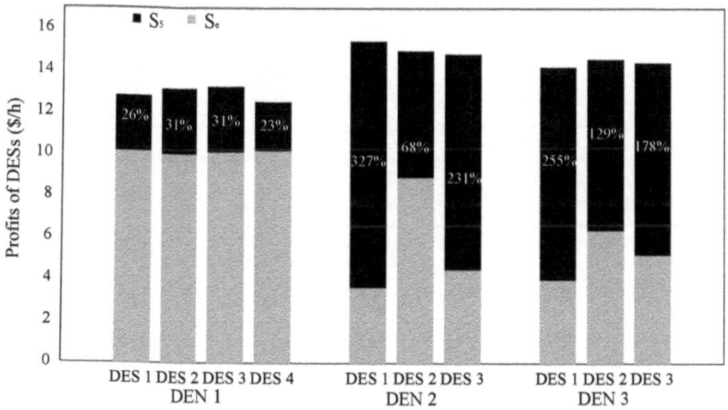

Fig. 6.22 Comparison of DESs' profits as for S_5 and S_e

enables the ISO to fairly treat the benefits of the E&G networks without being partial to any single utility network while forcing the other one to suffer huge losses.

As for DESs, Figs. 6.22, 6.23, and 6.24 illustrate the comparison results of DESs' profits between S_5 and the other three, respectively. Table 6.12 shows the wholesale prices and the retail prices of energies in these four cases. It can be seen from the figures that all the profits of DESs obtained by S_5 are significantly higher than those obtained by the other three solutions. From Tables 6.10 to 6.12, we can see that as for solutions S_e and S_{eg}, all the amount of natural gas traded with DESs are much larger than those in solution S_5, which incurs higher LMPs of natural gas. Similarly, singly minimizing the operating cost of natural gas network (S_g) results in larger amounts of electrical energy traded as well as higher LMPs of electricity compared with those of S_5. Therefore, it is sufficient enough to argue that the wholesale price will increase with the increase of the amount of energy traded. However, a high amount of energy traded accompanying with a high wholesale price doubly increase the generation cost of each DES, which will inevitably do great harm to the benefit of each of them. Although S_{eg} can effectively achieve the social optima, each DES surfers great loss and it is not envy-free (Tushar et al. 2015) so that those who are treated unfairly will definitely reject to accept this solution. Therefore, we can conclude that the strategies in S_{eg}, S_e or S_g fail to fairly treat all participants in the IES. However, in S_5, coordinating the benefits of the E&G networks can reasonably determine the interactive variables between utility networks and DESs, and balance the trading amounts and the wholesale prices of E&G. By doing so, the cost of each DES can be reduced, which will eventually bring about a harmony effect on increasing the profit of each DES. Furthermore, in S_5, all the retail prices of electricity are lower than the wholesale prices, which enhances the economic efficiency of the IES. Therefore, the idea of adopting the HMOGTA can balance not only the benefits of E&G networks but also the trading amounts and the wholesale prices of E&G, by which fairly treating various kinds of participants with different interests can be successfully achieved.

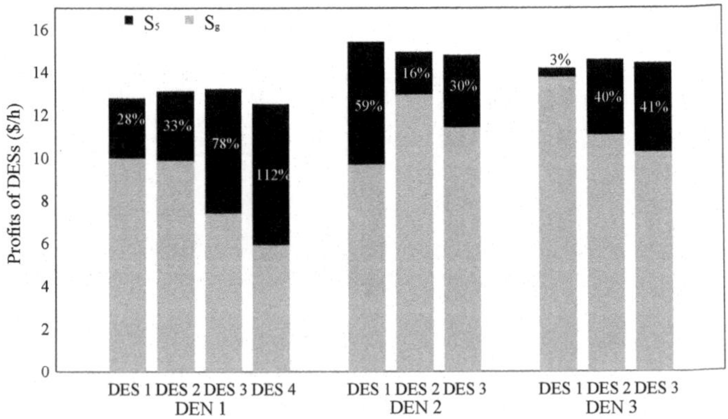

Fig. 6.23 Comparison of DESs' profits as for S_5 and S_g

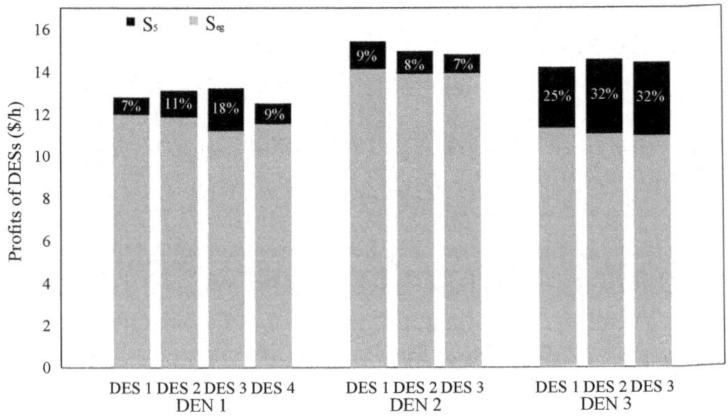

Fig. 6.24 Comparison of DESs' profits as for S_5 and S_{eg}

This section has proposed a hybrid multi-objective optimization and game-theoretic approach (HMOGTA) to solve the optimal operation of the IES. A Stackelberg game model has been developed to analyze the equilibrium strategies of DESs and EUs in each DEN. The Stackelberg Equilibrium of the game has been obtained in a closed form and the CDCENs of DESs have been derived based on the results of the game. The CDCENs, which are composed of the parameters of both the EUs' demand characteristics and DESs' operating characteristics, indicate the correlation between the wholesale prices and DESs' demands of natural gas and electricity. The CDCENs depict the interactions between utility networks and DESs, and demonstrate the substitutive relationship between the amount of E&G purchased by DESs.

The models developed for the IES can be tackled by the proposed HMOGTA, considering the operational constraints and the individual benefit of each participant in the IES. Simulation results have manifested the conflicting relationship between the operating costs of the E&G networks with consideration of the CDCENs. The

HMOGTA has successfully avoided partial treatment to any single individual while forcing the others to suffer huge losses. Moreover, using multi-objective optimization to solve the coordinated scheduling of E&G networks subject to the CDCENs not only has fairly treated the benefits of these two utility networks but also has brought about higher profits of all DESs, with the highest one up to 327%. The proposed HMOGTA reasonably determines the interactive variables between the participants and fairly treats the benefits of electricity network, gas network, DESs, and EUs in the IES.

The proposed method promotes the active participation of various kinds of participants with different objectives in the IES, by which the fair treatment for all individual participants can be ensured.

6.4 Summary

This chapter presents the study of coordinated scheduling strategy to optimize conflicting benefits for daily operation of the integrated energy systems with electricity and gas networks embedded. First, the models developed for the electricity and gas networks can be tackled by the proposed CSS, considering the operation constraints and the competing benefits of the two networks. Compared with the single-objective optimization for the total operation cost, the CSS considers the operation costs of the electricity network and gas network coordinately, and searches for the optimal solutions for the two networks. Moreover, according to the simulation results of sensitive analyses, the interdependency of the electricity network and gas network is related to the variation of the heating and cooling load. Instead, the electricity load and gas load only have an influence on their own network, but a tiny impact on the interdependency between the two networks. On the other hand, the number of electricity-driven compressors has an influence on the interdependency between the gas load variation and the power generation of the electricity network.

References

Amir R, Grilo I, Jin J (1999) Demand-induced endogenous price leadership. Int Game Theory Rev 1(03n04):219–240

An S, Li Q, Gedra TW (2003) Natural gas and electricity optimal power flow. In: Transmission and distribution conference and exposition, 2003 IEEE PES, vol 1. IEEE, pp 138–143

Auger A, Hansen N (2012) Tutorial CMA-ES: evolution strategies and covariance matrix adaptation. In: GECCO (Companion), pp 827–848

Bai L, Li F, Cui H, Jiang T, Sun H, Zhu J (2015) Interval optimization based operating strategy for gas-electricity integrated energy systems considering demand response and wind uncertainty. Appl Energy 1(1):1–10

Balcombe P, Rigby D, Azapagic A (2015) Environmental impacts of microgeneration: integrating solar PV, stirling engine CHP and battery storage. Appl Energy 139:245–259

Chaudry M, Jenkins N, Qadrdan M, Wu J (2014) Combined gas and electricity network expansion planning. Appl Energy 113:1171–1187
De Wolf D, Smeers Y (2000) The gas transmission problem solved by an extension of the simplex algorithm. Manag Sci 46(11):1454–1465
Deb K, Pratap A, Agarwal S, Meyarivan T (2002) A fast and elitist multiobjective genetic algorithm: NSGA-II. IEEE Trans Evol Comput 6(2):182–197
Dommel HW, Tinney WF (1968) Optimal power flow solutions. IEEE Trans Power Appar Syst 10:1866–1876
EIA-US (2011) Annual energy review. Energy information administration, US department of energy, Washington, DC. http://www.eia.doe.gov/emeu/aer
U.S. Energy Information Administration, Monthly energy review. https://www.eia.gov/totalenergy/data/monthly/pdf/mer.pdf
Gebremedhin A (2012) Introducing district heating in a Norwegian town-potential for reduced local and global emissions. Appl Energy 95:300–304
Gil M, Dueñas P, Reneses J (2016) Electricity and natural gas interdependency: comparison of two methodologies for coupling large market models within the European regulatory framework. IEEE Trans Power Syst 31(1):361–369
He S, Wu QH, Saunders J (2009) Group search optimizer: an optimization algorithm inspired by animal searching behavior. IEEE Trans Evol Comput 13(5):973–990
Hu Y, Bie ZH, Ding T, Lin YL (2016) An NSGA-II based multi-objective optimization for combined gas and electricity network expansion planning. Appl Energy 167:280–293
Jing ZX, Jiang XS, Wu QH, Tang WH, Hua B (2014) Modelling and optimal operation of a small-scale integrated energy based district heating and cooling system. Energy 73:399–415
Li YZ, Wu QH, Li MS, Zhan JP (2014) Mean-variance model for power system economic dispatch with wind power integrated. Energy 72:510–520
Li YZ, Wu QH, Jiang L, Yang JB, Xu DL (2016) Optimal power system dispatch with wind power integrated using nonlinear interval optimization and evidential reasoning approach. IEEE Trans Power Syst 31(3):2246–2254
Liu C, Shahidehpour M, Fu Y, Li Z (2009) Security-constrained unit commitment with natural gas transmission constraints. IEEE Trans Power Syst 24(3):1523–1536
Liu X, Wu J, Jenkins N, Bagdanavicius A (2016) Combined analysis of electricity and heat networks. Appl Energy 162:1238–1250
Martínez-Mares A, Fuerte-Esquivel CR (2012) A unified gas and power flow analysis in natural gas and electricity coupled networks. IEEE Trans Power Syst 27(4):2156–2166
Milgrom P, Shannon C (1994) Monotone comparative statics. Econometrica: Journal of the Econometric Society, pp 157–180
Mokhatab S, Poe WA (2012) Handbook of natural gas transmission and processing. Gulf Professional Publishing
Niknam T, Khodaei A, Fallahi F (2009) A new decomposition approach for the thermal unit commitment problem. Appl Energy 86(9):1667–1674
Niknam T, Azizipanah-Abarghooee R, Aghaei J (2013) A new modified teaching-learning algorithm for reserve constrained dynamic economic dispatch. IEEE Trans Power Syst 28(2):749–763
Overbye TJ, Cheng X, Sun Y (2004) A comparison of the AC and DC power flow models for LMP calculations. In: Proceedings of the 37th annual Hawaii international conference on system sciences, p 9
Partovi F, Nikzad M, Mozafari B, Ranjbar AM (2011) A stochastic security approach to energy and spinning reserve scheduling considering demand response program. Energy 36(5):3130–3137
Qadrdan M, Wu J, Jenkins N, Ekanayake J (2014) Operating strategies for a GB integrated gas and electricity network considering the uncertainty in wind power forecasts. IEEE Trans Sustain Energy 5(1):128–138
Qiao Z, Guo Q, Sun H, Pan Z, Liu Y, Xiong W (2016) An interval gas flow analysis in natural gas and electricity coupled networks considering the uncertainty of wind power. Appl Energy (in press)

Qiu J, Dong ZY, Zhao JH et al (2015) Multi-stage flexible expansion co-planning under uncertainties in a combined electricity and gas market. IEEE Trans Power Syst 30(4):2119–2129

Salimi M, Ghasemi H, Adelpour M, Vaez-ZAdeh S (2015) Optimal planning of energy hubs in interconnected energy systems: a case study for natural gas and electricity. IET Gener Transm Distrib 9(8):695–707

Schroeder JR, Donald W, et al (2010) A tutorial on pipe flow equations. In: PSIG annual meeting, Pipeline simulation interest group, p 1

Street A, Barroso LA, Chabar R, Mendes AT, Pereira MV (2008) Pricing flexible natural gas supply contracts under uncertainty in hydrothermal markets. IEEE Trans Power Syst 23(3):1009–1017

Strogatz SH (2014) Nonlinear dynamics and chaos: with applications to physics, biology, chemistry, and engineering. Westview Press

Talatahari S, Azar BF, Sheikholeslami R, Gandomi A (2012) Imperialist competitive algorithm combined with chaos for global optimization. Commun Nonlinear Sci Numer Simul 17:1312–1319

Tan ZF, Zhang HJ, Shi QS, Song YH, Ju LW (2014) Multi-objective operation optimization and evaluation of large-scale NG distributed energy system driven by gas-steam combined cycle in China. Energy Build 76:572–587

Tang WH, Spurgeon K, Wu QH, Richardson ZJ (2004) An evidential reasoning approach to transformer condition assessments. IEEE Trans Power Deliv 19(4):1696–1703

Thapar V, Agnihotri G, Sethi VK (2011) Critical analysis of methods for mathematical modelling of wind turbines. Renew Energy 36:3166–3177

Tichler R, Lehner M, Steinmüller H, Koppe M (2014) Power-to-gas: technology and business models. In: Briefs in Energy. Springer

Ting T, Rao M, Loo C (2006) A novel approach for unit commitment problem via an effective hybrid particle swarm optimization. IEEE Trans Power Syst 21(1):411–418

Topkis DM (1978) Minimizing a submodular function on a lattice. Oper Res 26(2):305–321

Topkis DM (1979) Equilibrium points in nonzero-sum n-person submodular games. SIAM J Control Optim 17(6):773–787

Tushar W, Zhang JA, Smith DB, Poor HV, Thiébaux S (2014) Prioritizing consumers in smart grid: a game theoretic approach. IEEE Trans Smart Grid 5(3):1429–1438

Tushar W, Yuen C, Smith DB, Poor HV (2015) Price discrimination for energy trading in smart grid: a game theoretic approach 99:1–12

Ugranli F, Karatepe E (2016) Transmission expansion planning for wind turbine integrated power systems considering contingency. IEEE Trans Power Syst 31(2):1476–1485

Üster H, Dilaveroğlu S (2014) Optimization for design and operation of natural gas transmission networks. Appl Energy 133:56–69

Varadarajan M, Swarup KS (2008) Solving multi-objective optimal power flow using differential evolution. IET Gener Transm Distrib 2(5):720–730

Venkatesh P, Gnanadass R, Padhy NP (2003) Comparison and application of evolutionary programming techniques to combined economic emission dispatch with line flow constraints. IEEE Trans Power Syst 18(2):688–697

Viana EM, de Oliveira EJ, Martins N, Pereira JLR, de Oliveira LW (2013) An optimal power flow function to aid restoration studies of long transmission segments. IEEE Trans Power Syst 28(1):121–129

Viswanathan GM, Buldyrev SV, Havlin S, Luz MGED, Raposo EP, Stanley HE (1999) Optimizing the success of random searches. Nature 401(6756):911–914

Wang JL, Wu JY, Zheng CY (2014) Simulation and evaluation of a CCHP system with exhaust gas deep-recovery and thermoelectric generator. Energy Convers Manag 86:992–1000

Wang JX, Zhong HW et al (2017) Review and prospect of integrated demand response in the multi-energy system. Appl Energy 202:772–782

Wang LX, Jing ZX, Zheng JH, Wu QH, Wei F (2018a) Decentralized optimization of coordinated electrical and thermal generations in hierarchical integrated energy systems considering competitive individuals. Energy 158:607–622

Wang K, Qi XX, Liu HD, Song JK (2018b) Deep belief network based k-means cluster approach for short-term wind power forecasting. Energy 165:840–852

Wang LF, Singh CN (2008) Balancing risk and cost in fuzzy economic dispatch including wind power penetration based on particle swarm optimization. Electr Power Syst Res 78(8):1361–1368

Wang MG, Tian LX, Du RJ (2016) Research on the interaction patterns among the global crude oil import dependency countries: a complex network approach. Appl Energy 180:779–791

Wang Y, Xia Q, Kang CQ (2011) Unit commitment with volatile node injections by using interval optimization. IEEE Trans Power Syst 26:1705–1713

Wang H, Yao X (2016) Objective reduction based on nonlinear correlation information entropy. Soft Comput 20(6):2393–2407

Wang H, Yao, X (2015) Objective reduction based on nonlinear correlation information entropy. Soft Comput: 1–15

Wang H, Yin W, Abdollahi E, Lahdelma R, Jiao W (2015) Modelling and optimization of CHP based district heating system with renewable energy production and energy storage. Appl Energy 159:401–421

Wei F, Jing ZX, Wu PZ, Wu QH (2017) A stackelberg game approach for multiple energies trading in integrated energy systems. Appl Energy 200:315–329

Wei F, Wu Q, Jing Z, Chen J, Zhou X (2016) Optimal unit sizing for small-scale integrated energy systems using multi-objective interval optimization and evidential reasoning approach. Energy 111:933–946

Wood AJ, Wollenberg BF (2012) Power generation, operation, and control. Wiley

Wu QH, Liao HL (2013) Function optimisation by learning automata. Inf Sci 220:379–398

Wu QH, Lu Z, Li MS, Ji TY (2008) Optimal placement of facts devices by a group search optimizer with multiple producer. In: IEEE Congress on Evolutionary Computation, 2008. CEC 2008 (IEEE World Congress on Computational Intelligence). IEEE, pp 1033–1039

Wu T, Yang Q, Bao ZJ, Yan WJ (2013) Coordinated energy dispatching in microgrid with wind power generation and plug-in electric vehicles. IEEE Trans Smart Grid 4(3):1453–1463

Wu QH, Zheng JH, Jing ZX (2015) Coordinated scheduling of energy resources for distributed DHCs in an integrated energy grid. CSEE J Power Energy Syst 1(1):95–103

Wu L, Shahidehpour M, Li ZY (2012) Comparison of scenario-based and interval optimization approaches to stochastic SCUC. IEEE Trans Power Syst 27:913–921

Xiong HG, Cheng HZ, Li HY (2008) Optimal reactive power flow incorporating static voltage stability based on multi-objective adaptive immune algorithm. Energy Convers Manag 49(5):1175–1181

Xu DH, Qu M (2013) Energy, environmental, and economic evaluation of a CCHP system for a data center based on operational data. Energy Build 67:176–186

Xu JZ, Sui J, Li BY, Yang ML (2010) Research, development and the prospect of combined cooling, heating, and power systems. Energy 35(11):4361–4367

Xu X, Jia HJ, Chiang HD, Yu DC, Wang D (2015) Dynamic modeling and interaction of hybrid natural gas and electricity supply system in microgrid. IEEE Trans Power Syst 30(3):1212–1221

Yang XS (2010) Firefly algorithm, levy flights and global optimization. In: Research and development in intelligent systems XXVI, pp 209–218. Springer

Yang JB, Xu DL (2002) On the evidential reasoning algorithm for multiple attribute decision analysis under uncertainty. IEEE Trans Syst Man Cybern Part A Syst Hum 32(3):289–304

Yang L, Jian J, Zhu Y, Dong Z (2015) Tight relaxation method for unit commitment problem using reformulation and lift-and-project. IEEE Trans Power Syst 30:13–23

Yao W, Wen JY, Cheng SJ, Jiang L (2012) Development of a matlab/simulink based power system simulation toolbox. Power Syst Technol 36(6):95–101

Yuan ZX, Jing ZX, Hu RX, Wu QH (2015) Operation optimization of CCHP-type microgrid considering units' part-load characteristics. In: Smart Grid Technologies—Asia (ISGT ASIA), 2015 IEEE Innovative, pp 1–7

Zeng Q, Zhang BH, Fang JK, Chen Z (2017) A bi-level programming for multistage co-expansion planning of the integrated gas and electricity system. Appl Energy 200:192–203

Zhai XQ, Wang RZ, Dai YJ, Wu JY, Xu YX, Ma Q (2007) Solar integrated energy system for a green building. Energy Build 39(8):985–993

Zhan ZH, Zhang J, Li Y, Chung HH (2009) Adaptive particle swarm optimization. IEEE Trans Syst Man Cybern Part B Cybern 39(6):1362–1381

Zhang X, Shahidehpour M, Alabdulwahab AS, Abusorrah A (2015) Security-constrained co-optimization planning of electricity and natural gas transportation infrastructures. IEEE Trans Power Syst 30(6):2984–2993

Zhang X, Shahidehpour M, Alabdulwahab A, Abusorrah A (2016) Hourly electricity demand response in the stochastic day-ahead scheduling of coordinated electricity and natural gas networks. IEEE Trans Power Syst 31(1):592–601

Zheng JH, Chen JJ, Wu QH, Jing ZX (2015) Multi-objective optimization and decision making for power dispatch of a large-scale integrated energy system with distributed DHCs embedded. Appl Energy 154:369–379

Zheng JH, Wu QH, Jing ZX (2017) Coordinated scheduling strategy to optimize conflicting benefits for daily operation of integrated electricity and gas networks. Appl Energy 192:370–381

Zhou Z, Liu P, Li Z, Pistikopoulos EN, Georgiadis MC (2013) Impacts of equipment off-design characteristics on the optimal design and operation of combined cooling, heating and power systems. Comput Chem Eng 48:40–47

Printed by Printforce, the Netherlands